Hafren
The Wisdom of the River Severn

Hafren
The Wisdom of the River Severn

Sarah Siân Chave

Illustrations by Rachel Elinor Collis

2025

© Sarah Siân Chave, 2025

All rights reserved. No part of this book may be reproduced in any material form (including photocopying or storing it in any medium by electronic means and whether or not transiently or incidentally to some other use of this publication) without the written permission of the copyright owner. Applications for the copyright owner's written permission to reproduce any part of this publication should be addressed to Calon, University Registry, King Edward VII Avenue, Cardiff CF10 3NS.

www.uwp.co.uk

British Library Cataloguing-in-Publication Data
A catalogue record for this book is available from the British Library.

ISBN: 978-1-915279-82-8

The right of Sarah Siân Chave to be identified as author of this work has been asserted in accordance with sections 77 and 79 of the Copyright, Designs and Patents Act 1988.
For GPSR enquiries please contact: Easy Access System Europe Oü, 16879218. Mustam.e tee 50, 10621, Tallinn, Estonia.
gpsr.requests@easproject.com

Cover design by Andy Ward
Illustrations © Rachel Elinor Collis
Typeset by Agnes Graves
Printed and bound by CPI (UK) Ltd, Croydon CR0 4YY
The publisher acknowledges the financial support of the Books Council of Wales.

For my sister,
Rachel
And for sisters everywhere

Contents

ix	List of illustrations
xi	Foreword
xiii	Map of Hafren
1	Opening
6	Chapter 1: Seeking the source Pumlumon, Powys
17	Chapter 2: Growing from Llanidloes, Powys, to Caersws, Powys
35	Chapter 3: Powering Newtown, Powys
50	Chapter 4: Bordering and crossing from Rhyd Chwima, Powys, to Shrewsbury, Shropshire
65	Chapter 5: Meandering Shrewsbury, Shropshire
78	Chapter 6: Industrialising from Shrewsbury, Shropshire, to Ironbridge, Shropshire
89	Chapter 7: Caring from Ironbridge, Shropshire, to Weston-under-Lizard, Shropshire
101	Chapter 8: Carving new routes Weston-under-Lizard, Shropshire, and beyond

112	**Chapter 9: Travelling onwards** from Upton Warren, Worcestershire, to Deerhurst, Gloucestershire
129	**Chapter 10: Feeling the pull of the tide, the pull of home** from Gloucester, Gloucestershire, to Oldbury-on-Severn, South Gloucestershire
147	**Chapter 11: Reuniting** Minsterworth, Gloucestershire, to Aust, South Gloucestershire, and Chepstow, Monmouthshire
163	**Reflecting**
166	**Acknowledgements**
168	**Endnotes**

List of illustrations

Chapter 1: Seeking the source
Hafren's source

Chapter 2: Growing
A linnet

Chapter 3: Powering
Weaving workshop, Newtown Textile Museum

Chapter 4: Bordering and crossing
Rhyd Chwima (Swift Ford)

Chapter 5: Meandering
The Welsh Bridge and Hafren seen through the *Quantum Leap* sculpture at Shrewsbury

Chapter 6: Industrialising
The Iron Bridge

Chapter 7: Caring
Turkey tail fungi

Chapter 8: Carving new routes
School at Weston-under-Lizard

Chapter 9: Travelling onwards
Twaite shad and salmon swimming

Chapter 10: Feeling the pull of the tide, the pull of home
Crane at Slimbridge

Chapter 11: Reuniting
Aerial view of Gwy and Hafren joining at Chepstow

Foreword

There's a marvellous spot at the human-made Diglis Island, just a short kingfisher flight away from the cathedral in Worcester, where you can descend some steel steps to visit the bed of the river and an underwater viewing gallery that allows views into a fish pass.

It's all glass and floaty light, as a 2 m × 2.5 m (6.5 ft × 8 ft) window provides a lens into the River Severn, with the opportunity to spot wild fish swimming by, silvery and determined. It's a privileged, magical view into what you could think of as a living aquarium, the huge fish pass working to slow down the water and provide a shallower gradient for migratory fish. If you stand there long enough, utterly enthralled, as I was, by the sight of fish rippling and muscling upstream, you might catch sight of as many as twenty-five species.

They make for a silvery, living list, from which you could easily assemble a found poem, conjured up from the watery depths. Bleak and bream and barbel, the latter so named because of little, distinctive features around their mouths, the Latin *barbula* meaning 'little beards'.

There are long-haul migrants such as salmon, of course, and dapper dace and stocky chub, finning through the water. See roach and rudd; slinky, sinuous eels like underwatery serpents, their bodies like one big muscle, as well as carp coasting through.

Tiny minnows, three-spined sticklebacks and bullheads pass by the glass, some at speed and some more leisurely in their progress. Occasionally, a big pike might submarine by, with what the poet Duncan Bush described as the 'neurasthenic grin of the Hollywood killer'. Busy tag teams of tench, plump trout and grayling; river lampreys and sea lampreys, each of these last species like something out of a science fiction movie: the suckers, the sharp-toothed stuff of nightmares. And to round off the list, a stone loach, maybe, or, if compiling a piscine A to Z, then register the allis shad and zander scything past.

But it's the twaite shad that make this place really important, as the weirs built in the 1840s stopped the fish from migrating further upriver. But the fish passes now allow them safe passage for the first time in almost 180 years.

This window on the underworld of the river is akin to the experience of reading *Hafren*, which follows the meandering course of the Severn, offering plentiful glimpses of real insight and plenty of riverine depth. Sarah Chave visits Diglis Island and so, so many other fascinating places, writing about them beautifully, while ever alert to the problems faced by the Severn and so many other lovely, long watercourses.

Chave sets out to learn from the river and to share the wisdom she imparts and to follow her musics, too, from the 'dance of lively high tones' near the peaty source to the unexpected roar of the river when the distinctive Severn bore is 'the sound of the sea brought inland'.

This is a book written with care and about care for the world we share, quietly urging us to look after things much better and to listen to the river's songs with more attention and thus appreciate what she has to teach us.

Chave has written a book that is nothing less than embracing, enfolding the reader into her words and stories even as she proves to be the best kind of a companion for a long walk – well-informed, curious and chatty, but also willing to pause awhile, to stand still, to fully appreciate the river's flowing life, brought fully into being by the bright splash of rain.

Jon Gower

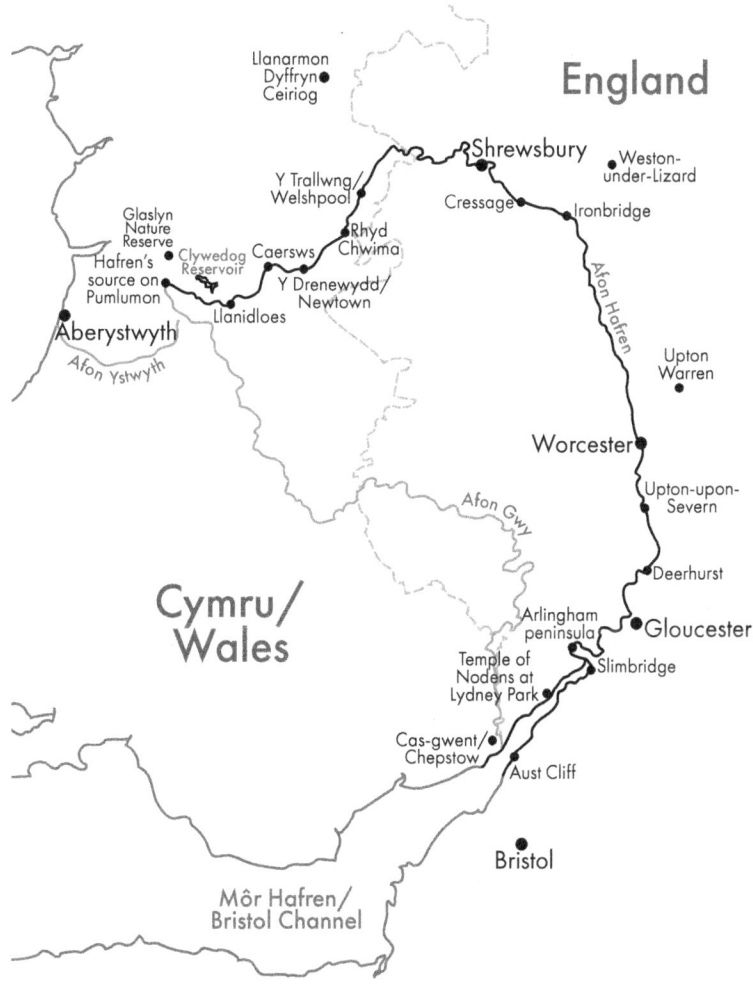

Opening

In Celtic myth, a giant sleeps under Pumlumon, the highest plateau in the Cambrian Mountains just inland from Aberystwyth in mid Wales. The story begins when the giant awakes from his slumbers and surveys the land around him. It is beautiful and, at first, he is happy. Yet he senses something is missing. There are no streams or rivers running through the landscape. He tries to think of ways he can bring water, but he cannot. Sadness overwhelms him, and three tears roll down his cheeks. As these fall to the ground, each tear becomes a pool with its own water nymph – the giant's daughters. He names them Hafren, Gwy and Ystwyth. For many years he nurtures these daughters, and their pools grow.

One day, he accepts that it is time for them to leave home and fulfil their destiny to join the sea. He calls his daughters to him and tells them about his decision. Ystwyth steps forward first. She is the smallest sister, always in a hurry and quick to make decisions: 'I will go west and reach the sea as soon as possible. I yearn to see it, smell the salt, feel the waves.' The giant grants her wish, and Ystwyth sets off, tumbling westwards for twenty miles, fulfilling her destiny by joining the sea at Aberystwyth. Hafren (translated as 'Sabrina' in Latin and 'Severn' in English) steps forward next: 'I want to fulfil my destiny, but I am not in a hurry. I want to see the lands and peoples and places along the way. I will be kind and bring them water, but if they abuse my gifts, I will rise up against them and show them my power.' The giant grants her wish, and Hafren sets off on her lengthy, meandering route through many towns and villages in Wales and England, finally completing her purpose of joining the sea near Bristol and Chepstow on the English–Welsh border. The last daughter to step forward is Gwy ('Wye' in English): 'I, too, am in no hurry, but I want to see natural beauty, not towns and people, before I fulfil my destiny.' Her father grants her wish, and Gwy sets off, travelling south through the beautiful hills and valleys of Wales, joining her sister Hafren near Chepstow before they enter the sea together. The giant is sad to see his daughters leave but, comforted by the knowledge that they will come back to him one day, he returns to his slumbers under Pumlumon.

This myth has much wisdom to share. It speaks of care, of family, of leaving and returning home. It tells of the long water cycle and how it brings water from the sea to the land. When water on the surface of the sea warms, it evaporates and rises upwards as vapour. As it rises, the vapour cools, and turns into water droplets, which form into clouds. These then release the droplets as rain, especially over high land near the coast, at places such as Pumlumon. A drop of water may pass over 3,000 years in the ocean but, on average, it lasts only nine days in the atmosphere before falling back to earth. Three metres (10 ft) of rain can fall on the Cambrian Mountains each year. Water eventually finds its way back to the sea through streams and rivers, and, once again, takes part in the long water cycle, falling as rain. The giant in the 'Three Rivers Myth' could indeed believe that one day his daughters would 'return' to him.

He might have been less able to accept our current era of climate change and ecological destruction and loss. Climate change is having a severe impact on the long water cycle. Warmer temperatures and rising sea levels caused by global warming are increasing rates of evaporation and precipitation. Worldwide, higher rainfall in coastal areas and increased drought and desertification inland are already being seen. More caring ways of being and acting in the world are urgently needed: ways that recognise that humans are an entwined part of nature, rather than something separate from it.

Opening

But how can humans, particularly those with mindsets in which 'nature' is understood as a 'resource' to be used, become more open to other ways to be in the world?

Celtic and other traditions provide a way forwards. In these ways of thinking, the wider natural world is a lively, vibrant participant with capacity to act and, with wisdom, to share. For example, in Celtic traditions, sources of water provide wisdom, and streams and rivers are a way that such wisdom is spread and communicated. But we have to be open to what they are communicating, pause, spend time with and listen to the water. We can also listen and be open to poetry, songs, histories and myths connected with the water, such as the giant's story. In another story, which hovers between fact and myth, we learn about Hafren (who was the daughter of Locrinus, the legendary king of England). She was drowned in the river by her stepmother Gwendolyn. As Hafren descended under the water, the river spirits took pity on the innocent girl and transformed her into a river nymph. She lives in the river to this day, reaching out to the world in the dewy mists that rise from her waters at dawn.

My idea and desire to write this book came during an online talk. An image suddenly filled my screen: a map of the UK and Ireland by Robert Szucs depicting the rivers and their tributaries as vibrant, flowing arteries, the lifeblood of the land. Hafren drew me in. Amidst all the rivers and details, I could immediately trace her route, feel her pulse, recall many memories and feelings linked to her. I grew up near Hafren, on the Staffordshire–Shropshire border. We heard of her ebbs and flows, her power to flood. We took our visitors to historic places she had helped to form. We were proud of her. Later, as an adult living in Devon, crossing her was an important point in our journey to visit my mother and sister in Wales. Her wide expanse, her sparkling waters and glistening mudflats, the flashes of green and purple as the colours of the bridge melded with the surroundings, and the '*Croeso i Gymru*' ('Welcome to Wales') sign, all marked moments of excitement and awe. I now, once again, live near Hafren, still on the English side. The restrictions on travel during the Covid-19 pandemic re-emphasised her role as a border. I spent many hours in South Gloucestershire looking across her to the distant hills of Wales: a place we could not visit, something I never thought I would experience in my lifetime. I had time to feel the breeze from the estuary, to reflect. I felt comforted by Hafren. She gave me a sense of scale: a sense that this was but a moment in her long history, a moment that would pass.

I decided to set off on an odyssey from her source to the sea to experience, record and learn from the wisdom I know she has to share. We live at a time where we need to be open to the wisdom of rivers; find new ways to live with and care for them and allow them to care for us, and through these relationships find new, more harmonious ways to live together with all (animals and plants, rivers and rocks...) on our shared planet. My hope is that on this journey I will become receptive to Hafren in ways that go beyond facts and figures – although these, too, are important and will be sought out. They give us additional insights, contexts for our experiences and understandings of Hafren's connection with deep time. They are also impressive in themselves. Hafren's source is at a height of 600 m (2,000 ft) above sea level, and she is the longest river in mainland Britain, travelling 220 miles (354 km). Today, her course is semicircular, initially north-easterly towards the Welsh border then southerly through England. However, her original route, many millennia ago, continued north-eastwards, joining the River Dee then draining into the Irish Sea near Liverpool. Ice and rocks in the Pleistocene Epoch (believed to have spanned the period 2.4 million to 12,000 years BCE) blocked her course near what is now the town of Shrewsbury. This forced her to make a dramatic change of route southwards, firstly through the narrow gorge at Ironbridge, eventually joining the Môr Hafren (Bristol Channel), the route so familiar to me and others today.[1]

As I write this, I look to the journey ahead. I know I will need the support and encouragement of family and friends, especially my son, James, and my husband, Peter. I do not have robust health and

Opening

cannot walk long distances, but I still want to experience a journey such as this. Together, we will seek ways to make this possible. I believe that everyone in different ways can spend time in the wider natural world, and experience all it offers to us, and each of these ways is valid and valuable. For this journey, we will first seek Hafren's source, her starting point, and then make different day trips and longer visits to follow her long, meandering route back to the sea. To listen to her wisdom, we will visit various places along her banks, spend time with her and with the people, birds, animals and plant life that live alongside her, as well as listening to stories, poetry and music linked to her and to these places. We will take occasional forays away from her banks to follow ideas that spring up along the way. The seasons will play their part in our journey, too. We will consider the amount of rain falling on the peatbogs, the times of fish and bird migration, the impact of flooding and the influence of high tides in the spring and autumn.

I hope you enjoy taking this journey along Hafren's course with us as well as being moved to remember, to spend time with and to learn from your own special rivers, whether the ones that have accompanied you through your life or ones you are yet to discover.

Chapter 1:
Seeking the source
Pumlumon, Powys

To reach Hafren's source I must first cross her broad estuary. My son James is beside me. I reach out to pat his arm and smile, happy and excited to be setting off together on this long-planned journey. The two bridges that link England to Wales near Bristol stand out, brilliant white, against the azure-blue spring sky. We take the lower, newer bridge. As we cross the wide expanse of water sparkling in the sunshine, I look towards the banks where the rivers Gwy (Wye) and Hafren (Severn) are reunited near Chepstow (Cas-gwent). I think of the many occasions I have travelled from my home in England over these bridges, or through the railway tunnel that runs under the water here. On today's journey, we continue west

Seeking the source

towards Newport. Then James turns the car northwards, broadly following Gwy's course towards the Pumlumon plateau in the Cambrian Mountains of mid Wales. As we travel between bracken-covered hilltops, green valleys and field after field of sheep and lambs, we catch occasional glimpses of Gwy. I tell James the giant's story and he says we will have to take care if we meet him on our walk. After we pass through the small market town of Llanidloes, the land begins to rise. Clouds build as we head into higher ground. I sit forwards, excited to be so close, to visit a place I have often traced on the map, but also nervous. Will we find the place to start our walk? Will it be too difficult? As we arrive at Hafren Forest car park, the starting point for walks along both Hafren and Gwy, the sky is grey, heavy with clouds, but we are undeterred. We pack our waterproofs and a picnic into James's rucksack. I know I am lucky to have him with me, and not just because he is carrying everything. Thus prepared, we set off into Hafren Forest, planted in the 1930s by the Forestry Commission.

The first part of the walk to Hafren's source is along a wooden accessible walkway set between the tall pines and the riverbank. I am struck by the pleasure Hafren is giving. She is about 2 m (6 ft) wide here, fast-flowing, still shallow. Two young children and their father are paddling across a ford. Other children are excitedly cycling along the walkway, their cheeks wobbling like jelly as they pedal over the uneven wooden planks. They laugh. We arrive at a picnic area next to a waterfall and chat with a group from Guatemala. One family now lives in Newtown and have brought their friends to enjoy all that Hafren and the forest offer. This is indeed a beautiful place. Willowy trees overhang the water that resonates as it cascades over the rocks, the varying notes and pitches blending tunefully with birdsong, happy voices and children's laughter. Textbooks tell us that waterfalls are formed as the different types of rock in riverbeds erode at different rates. Yet they are so much more. In Celtic traditions, waterfalls are places where the wisdom carried and communicated by rivers is intensified. Is that why humans are drawn to them still, even in modern Western cultures? Can we feel their pull? What are they saying to us? I find myself reflecting on the timelessness of bringing visitors to waterfalls and rivers – something we, too, had done when I was a child. The group asks us where we are from and why we are here. We tell them how only that morning we had crossed the mouth of the river as she entered the sea, and we are now on a quest to find her source. They wish us luck.

We set off again. A sign warns that the path to the source is 'strenuous'. At first, the way is level, running beside Hafren, already narrower – about 1 m (3 ft) wide. Smaller trees and bushes live on her banks amidst the lofty pines. We stop at a place where two waterways meet, Hafren and Hore, to consult our paper map: no internet here. We find our way and, as the path steepens, the scene around us becomes increasingly dominated by the commercial pine tree plantation. Hafren, now a stream, flows fast through a gully below us. We can see the open mountain top beyond the forest, and the myth of the giant sleeping beneath it comes vividly to mind. The dried stems of last year's heather, now a deep rust, are the giant's hair, maybe; surely a Welsh giant would have curly, red locks. Areas of forest clearance stand out starkly grey amidst the tall green growth. Beside the path, water drips through mossy outcrops, and we pause. I put my hands forwards and feel the water flow over them. It is cool. I wash my face, hot from the climb. 'Drink some,' says James. I put my head forwards, adjust my face upwards, open my mouth and feel the water running across my tongue – refreshing, mossy. I feel pleasure. I think of drinking from school water fountains as a child, the novel sensation of flowing water – its movement lost, contained, when we drink from a glass or bottle. Now it is James's turn: 'Mmm, earthy, lovely.'

We press on, upwards, towards the tree line at the edge of the forest. We pass a place where pines on the opposite bank have fallen, in strong winds maybe, lying higgledy-piggledy, unlike the neat areas of human-felled trees. 'The giant's matchsticks,' comments James, and we imagine the giant crashing through the forest here. In one spot, a fallen tree has created a bridge across Hafren, now only about 45 cm (1.5 ft) wide. Moss in many shades of green surrounds us. We step over stone-lined drainage channels. These are dry now. The sound of Hafren is constantly with us – melodic – a dance of lively high tones. We meet a family and ask them how much further it is to Hafren's source. 'Not too far,' they reply, 'and it's a bit easier than the section you have just done.' We admire the child's stick. 'Really helpful for climbing mountains,' I say. 'Yes, and for going down, too,' the child replies. They are from Aberaeron, a small town on the coast in Ceredigion, south of Aberystwyth. James and I know Aberaeron well. My mother's ashes are buried at Llanina (St Ina's Church) on a headland nearby. The family are surprised we know their town and are pleased to hear our enthusiasm for it. I feel connection: an unexpected and special moment of meeting and memories.

We break through the tree line and enter the open upland blanket bog. I crouch down to examine the bog more closely. Sphagnum mosses form spongy cushions amidst the tough grasses and heathers. Blanket bogs are ombrotrophic, which means that they receive water from precipitation rather than nutrient-rich springs or streams. The soil is, therefore, nutrient-poor and acidic, only able to support plants adapted to these conditions. When plants die, microorganisms in the soil cause the plant material to decay. In waterlogged areas such as these, only microorganisms adapted to live with the low levels of dissolved oxygen available in water can survive. These can only digest plant material through a *slow* process called anaerobic digestion.[2] The partly decomposed plant matter accumulates at a faster rate than it decomposes, and peat is formed, thereby storing underground the carbon that the plants had drawn into their structures during their lifetime.

We enter a gully carved by Hafren through the bog. Suddenly, we are surrounded by skylarks, their song spiralling upwards with their flight. They swoop down towards us, flying so close. Wow. I am filled with joy and turn my face upwards towards them. We stop to listen. Hafren's voice merges with their music. Skylarks and their song are a passion of mine, and I did not expect to hear them today. We walk through an area where mist surrounds us, as if we are in the clouds: a welcome coolness. The mist lifts and we move on. Hafren is a narrow stream now but still powerful, running fast. Amidst her high notes we can hear a rumbling. It is so distinct that we think at first it might be a far-off plane, but no, there is no plane in the sky. Maybe it is the giant gently snoring as he slumbers under this mountain? We clamber down the bank to investigate. The rumbling seems to be coming from areas where the bank overhangs Hafren, perhaps creating an echo.

We meet two young women. 'Not far now,' they reassure us. 'Mind you, it's pretty gross by the source. All boggy, rank. But do go on further, to the white stone cairn about 200 metres beyond the source,' they say. 'You can see the sea from there.' So, heartened, on we go. All around we see curved ridges of black-brown peat raised like giant eyebrows above the surrounding, lower land. I stand next to one and the exposed peat is as tall as me, indicating significant drainage and a reduction of the 'wateriness' needed for healthy peatlands. Human relationships with peatbogs are complex, and Western cultures have a tendency to be dismissive of them. Humans do have cause to be wary. We can become trapped in these boggy areas. The slowly occurring plant decay within them can produce noxious odours such as hydrogen sulphide. We are alert to danger. Is it an inbuilt instinct of self-preservation that causes people to find bogs 'gross' and 'rank'? Or is this a learnt response instilled in certain cultures?

Attitudes towards peatlands are, however, beginning to change. Whilst they only cover 3 per cent of the world's surface, they hold 3 billion tonnes – 30 per cent – of the world's stored carbon. This amazing statistic highlights the importance of protecting peatlands both for their own sake and to prevent the release of this carbon back into the atmosphere. The carbon is released when the semi-decomposed plant matter in peat is exposed to high levels of oxygen in the atmosphere, and decomposition (withering) is completed rapidly. The UK has a significant role in caring for these important habitats since 12 per cent of the world's total peatlands are in the UK and, of these, 90,000 hectares (222,395 acres) are in Wales (4.3 per cent of Wales's land area). UK peatlands store twenty times more carbon than all the UK forests. However, 80 per cent of the UK's peatlands are in need of restoration and it is estimated that only 10 per cent of the peatlands in Wales are in a healthy condition.

One metre (just over 3 ft) of peat takes 1,000 years to form but only moments to destroy. Such destruction occurs for many reasons. Drier peatlands caused by climate change are susceptible to fire that releases carbon from both living plants and the semi-decayed plant material stored underground. These underground fires are hard to put out and they can smoulder for extensive periods. Peatlands are susceptible to pollution, for example, from run-off from land treated with nitrogen fertilisers. The extra nutrients this introduces encourage growth of heathers that suppress the sphagnum mosses that have a tremendous capacity to absorb and retain the water critical for the healthy functioning of peatland. The loss of the mosses causes further drying out, adding to a vicious cycle of destruction. Over- and undergrazing also contribute to decline in the plant life needed for peatland health and the biodiversity it supports, as does the widespread introduction of conifer monoculture, such as the plantations we can see below us. This type of planting was particularly prevalent in the post-war period when timber production was prioritised over biodiversity. The practice of cutting peat for fuel does release stored carbon into the atmosphere. However, this traditional practice tends to be on a restricted scale. Far more widespread and concerning is the practice of extracting peat for horticultural use, although new UK legislation is now phasing this out. As we learn more about the crucial role peatbogs play for biodiversity and carbon storage, there is hope that they can be saved. For example, in November 2020 the Welsh Government funded a five-year peatland restoration programme led by Natural Resources Wales, and progress is being made.

Hafren

James and I press on with our walk. At last, we can see the post that marks the boggy area identified as Hafren's source. We quicken our pace, excited. We approach and see the dark pools of water, lower than the surrounding peatland, a walkway, several metres long and a marker post, nearly 2-m (6-ft) tall, set into a granite platform. We cross over, and I trace the words carved into the post: '*TARDDIAD AFON HAFREN*'[3] on one side, 'SOURCE OF THE SEVERN' on the other. A solitary bird of prey circles overhead, its sombre cry emphasising the loneliness of this place. The dark pools are edged with floating tendrils of pondweed and clumps of soft reeds sit like islands in the dark water. Puddles and small boggy pools have formed in the path, and the dark, peaty water reflects the surrounding plants and the fast-moving clouds. The surrounding ridges reveal the dark peat now exposed to the air. All around peatland extends for miles, an upland expanse of pale yellows and deeper reddy-browns: a place with its own beauty. It is here that rain falls onto the land, trickles into gullies and small hollows, flows through rivulets and forms into this boggy area: a watershed – a place of multiple beginnings. A change in gradient starts Hafren's trickle downwards, soon swelled by more water dripping through the mosses and grasses on her banks. How do we feel arriving here? Elated, thoughtful, grateful: a shared moment to treasure. The moment is tinged with sadness, too: sadness as we miss my husband Peter, not well enough to be with us today; sadness for the damage to this land evidenced by the exposed peat, and to land and its inhabitants everywhere.

Looking upwards, we can see the white cairn mentioned by the young women we had met earlier. Remembering their promise that we could see the sea from there, James is keen to walk on. As we do, he comments on the silence. We have grown used to hearing Hafren's song and now miss her presence. The story of the giant's melancholy at the lack of water flowing over his land comes to our minds. I pause by a lonely pool of water and think of the giant nurturing the pools that formed from the tears he shed: a story of care for this special landscape that we all need to heed. Our legs object, but the extra climb is well rewarded. At the white cairn we sit and can see the mountains unfold in all directions, and to the west, the distant coast. Over the sea there are many patches of blue sky, but as the land rises, clouds mass increasingly in dark layers, marching towards us. We feel a few specks of rain against our skin and moisture soaking through our clothes from the ground beneath us: the water cycle in all its glory. We can feel drizzle in the air, damp on our cheeks and curling our hair. We are suspended in a watery world.

We need to treasure these clouds, part of the long water cycle that, as the giant's story foretold, brings water to us from the ocean where 97 per cent of the hydrosphere (the term for all the water in the world) is stored. It is here, rather than the boggy area marked as Hafren's source, that we eat our picnic. Sitting here, by this cairn, with these layers of dark clouds approaching, these soft mosses beneath my fingers, the distant sea and Hafren's source nearby, I can reflect on the damage done to these peatlands. I can celebrate the increasing recognition of their importance and the actions being taken to restore and care for them. I can also feel *this* place, at *this* moment, touch the soft ground, hear the wind, feel it on my face, hear the cries of the birds. I can treasure *these* encounters with *these* others. Rather than approach them as 'objects of study', can I be open to them, approach them with tenderness, and humility: an entwining in what the philosopher Martin Buber calls an *I–Thou* reciprocal relationship that then alters who I am and what I do in the world? I hope so.

Evening begins to fall, and it is time to leave. We stop again at Hafren's source, crouching down to observe the small insects buzzing over the tendrils of pondweed. Faint glimmers of sunshine cast my shadow over the water in a moment that I commit to my mind's eye – to be drawn on in the future when I want to re-feel this special place. We linger at the point where water forms a narrow channel, and Hafren begins her journey downwards. We follow

Hafren

her, walking on her left bank, so-called as we have our backs to her source. The descent is steep in places. The young boy we met on the way up was right: a stick is helpful for going down as well as climbing mountains. Hafren soon grows from a narrow trickle to a lively, fast-flowing stream. All around us water is dripping through the mosses and peat in the sides of the gully she has formed, adding to her growing strength and breadth. Suddenly, a skylark in full song swoops low above our heads, so close, taking us by surprise. We leave the peatland behind and re-enter the forest, pausing at a waterfall where a rock, sculpted by erosion into a pyramid, perches at an angle over Hafren's bubbling waters. 'This is my favourite waterfall,' says James. It does have presence. We peer down into the deep, dark, green pool formed below the waterfall. 'So tempting for a swim,' he says. We must return in the summer.

As we walk through the forest, I tell James about a visit I had made, the year before with Peter, to Glaslyn lake and nature reserve, which is about 4 miles (6.4 km) north of here. The lake sits beside the old mountain road that winds its way between Llanidloes and Machynlleth. We had travelled there to see one of the restoration activities of the Pumlumon Project, established in 2007 and run in conjunction with the Montgomeryshire Wildlife Trust. The project aims to revive the ecology and economy of the Welsh uplands in partnership with local communities and to encourage high nature value farming. It has adopted a 'working-with' farmers and landowners approach, emphasising that it is the people who live and work on the land, often for generations, who are best placed, and most committed, to effect and maintain change. Restoration work has involved a shift away from a dominance of grass towards a mix of heather moorland, deep peats and sphagnum bogs and provides financial incentives to promote this.[4] Small diggers have been used to block one end of previously installed drainage channels. This has significantly raised the water level and led to rewetting of the bog as well as restoring the peat. These activities have also encouraged the growth of rare plants including stiff sedge, spring quillwort, dwarf willow, starry saxifrage and rare and scarce mosses and liverworts. Skylarks, merlins and red kites are making their homes there.[5] In contrast to today, here, near Hafren's source, Peter and I had not seen raised eyebrows of exposed peat at Glaslyn.

As we descend further into the forest, Hafren widens, and the clouds begin to clear. Feeling the warmth from the sun, we discuss putting our feet in the water before leaving. We hesitate as it seems an effort. We are tired but we decide we must, at least with one foot. Bathing in the river, a ritual recognised as sacred in so many world cultures, is one we feel the need to perform. The water is cold but oh-so-soothing on our weary feet. As we remove them, we feel warmth as our bloodstreams respond to Hafren's chill. It is so wonderful that we both decide to bathe our other foot as well. There is humour in all this, too, as we hop on one foot trying to dry the other on a spare pair of socks. It is late now, almost sunset, but approaching the ford near the car park we see there are two new children and their father playing, keen to enjoy the last moments of the day. We exchange greetings then reluctantly return to our car.

Setting off eastwards towards Llanidloes down the steep, narrow road, the sun is setting behind us. It lights up the valley ahead: truly the golden hour. The trees overhang the lane. Incongruously, we see a cat beside this lonely road, but then a farm appears, perched above Hafren, wider now, flowing rapidly over a rocky bed. We reflect on our day. For me, the highlight was the journey to the beginnings of Hafren, the river I have known all my life. It was listening to her changing voice, the notes high and low, so different as she ran over rocks, boomed under banks, cascaded into pools below waterfalls. It was hearing the birdsong, feeling the soft mosses, tasting her flowing water and bathing our feet. It was experiencing the multiplicity of her sources, the mist, the peatbogs, the water dripping through the mosses on her banks, the rivulets joining her. I recall the story told by some friends of searching for the source of the River Torridge. After several hours of squelching through boggy culm grassland dominated by a type of purple moor grass peculiar to north Devon and on hearing rivulets of water underground, they realised they needed to challenge their preconceived ideas. There was, perhaps, no single source; rather, there was opportunity for inhabiting a watery togetherness.

Humans, too, have a multiplicity of sources, with many intertwined origins and antecedents. Whilst we may search for one source, one reason for our place in life, we are, instead, called to seek out, explore and find connection with our many roots, life experiences and the emotions that accompany these. Being here, with Hafren, has given me time to explore my feelings rather than push them aside or bury them deep. Somehow, Hafren's presence, the melodies of the water and the birds, the rhythm of walking, the love and friendship of my son, have all helped me to process my emotions, both happy and sad, and make peace with them.

James enjoyed journeying up to the peatlands where Hafren begins, but, for him, it was following Hafren outwards that struck a chord. It was seeing the trickle emerge from the bog and head downstream. It was experiencing water from all around joining her fast-flowing waters: my son and Hafren, both growing, on the brink of new adventures. In Llanidloes, 10 miles (16 km) from her peatland origins, Hafren becomes an established river, swelled by her many tributaries. A bridge with three strong arches is needed to cross her. As I look down from the centre of the bridge into the clear, fast-flowing water, I take time to appreciate her boggy beginnings, the role that peatlands play in healthy ecosystems and all that Hafren has shared with us today.

Chapter 2:
Growing
from Llanidloes, Powys, to Caersws, Powys

Hafren arrives at Llanidloes as a stream and leaves as a river. The town's descriptively named 'Short Bridge' and 'Long Bridge' highlight how rain, groundwater and tributaries swell her waters both here and throughout her long journey to the sea. Summer has arrived and I have returned to Llanidloes, this time with Peter. Today, I am searching for Afon (River) Dulas and Afon (River) Brochan – known as headwater tributaries – as they join Hafren near her source. The two rivers flow into Hafren just before she enters the town. Looking over the side of the single-arched Short Bridge, I can see Hafren's dappled, gurgling waters below, flowing freely, no

Hafren

longer in service to the former flannel mill standing silently beside her. The sky is a beautiful azure blue with fluffy, picture-book clouds scudding across it. The sun is warm on my face as I set off up the lane towards Felindre. The houses on one side of the lane look across to an ivy-clad, low stone wall. Beyond this, ferns reach down towards the water, about 1.2 m (4 ft) wide here. As the lane rises, the houses and wall are replaced by hedges of willow, oak, alder and beech. The soft-spined cases around the beech-nuts are beginning to appear. I can see occasional glimpses of Hafren through the leaves, her water plashing over her stony bed, brightly lit and sparkling in places but darker and mysterious where the banks and trees overhang. Stony outcrops protruding from the banks are inviting – places to sit and cool my feet in the water – but there is no way to reach them. The walls and hedges are a solid deterrent.

   I walk uphill, then take the lane on the left that dips towards Felindre Bridge where I stop to look over the stone parapet. Even in midsummer, Hafren is fast-flowing, and I try to imagine the strength of her flow midwinter after heavy rain: power once harnessed by the mill here to grind corn in the nineteenth century. The mill house, upstream from the bridge, is now a private home. Looking downstream, I crane forwards, trying to glimpse where the tributaries join Hafren, but my view is blocked by another private house built on a bend. I look for a way down to the water, hoping to be able to paddle downstream to the place where the tributaries join Hafren, but to no avail. Crossing the bridge, I turn left and walk along the lane past some modern houses and onwards to a stile and footpath leading across a field towards Afon Brochan. I look up and enjoy the sunshine, the birdsong, the view up the gently sloping valley rising up toward Glyn-Brochan, the cows grazing, the thick hedges and clusters of trees. Consulting my map, I can see that Afon Brochan joins the larger Afon Dulas just across this field, before both flow into Hafren, so I set off. Intent on my destination, I fail to notice how the herd of about thirty cattle has approached and are now just in front of me. They are a mix of

Growing

black, brown and white bullocks, intently curious, keen to play but also somewhat intimidating. I glance towards Afon Brochan and can see the mud on her banks but, even on tiptoe, I can't see water. One white bullock approaches to within a few metres, particularly inquisitive. I look back to the lane and forwards towards Brochan's bank. Reluctant to place the herd between myself and the stile, I hesitate, then turn, and slowly head back. The herd soon lose interest, and I can feel in my body the vibration their hooves make as they thunder away. The single bullock is more persistent and follows me. I feel his intent gaze but, much to my relief, he, too, loses interest eventually, and I glance surreptitiously backwards to see him paw the ground one last time, turn and wander off. Today, it seems, is not destined to be the occasion I will see these tributaries and the place they join with Hafren. I feel a little disappointed, but I have seen the land through which they flow and, for now, this is enough. Strolling back towards Llanidloes, I make a final attempt to reach water and feel it flowing across my hands, my feet. An open gate gives access to a patch of nettle-dense ground. I pick my way across, nervously looking towards a house nearby. But as I approach Hafren, I see a wire fence hidden until now by trees. I crouch down beside it, happy to rest here for a moment, reflecting on how even though the water is fast-flowing, its presence is calming and soothing. I am content.

From the Short Bridge, I head towards the Long Bridge, passing Llanidloes's black-and-white-timbered old market hall constructed early in the 1600s, although the town is much older than that. It received its first charter in 1280 but probably dates back to the seventh century. The walk from Short Bridge to Long Bridge is not far, a matter of a few hundred metres, but the change in Hafren is striking. Now this significant bridge, supported by its three strong arches, is needed to cross her. The cause of this dramatic increase is another tributary – Afon Clywedog – which joins Hafren here. Opposite the confluence of these two rivers is a grassy area with benches. The first time I saw this tributary I was sitting with James enjoying the spring sunshine. It was hard to discern Clywedog's flow, and I was unsure I had actually seen the confluence, having to check later on a map that this is the point where the two rivers meet. Yet today, in the middle of a summer drought, there is no mistaking Clywedog's presence. Water is rapidly flowing from her, as though a tap has been left open. There is a reason for this. Her flow is managed by human hands, holding back and releasing water at the dam built across her higher up the Clywedog valley, which I had seen on the journey I had made to Hafren's source with James. Having made a wrong turn in Llanidloes, we had taken a somewhat circuitous route to Hafren Forest along a narrow rollercoaster-like moorland road. Cresting a high point, we had looked down at the top of the dam plunging a vertiginous 74 m (243 ft) from the glassy surface of the water to the valley below. The effect had been even more striking as we were not expecting it. Perhaps I have an overactive imagination, but I always find dams somewhat terrifying, imagining the huge quantities and power of the water trapped behind them and the water breaking through and flooding into the valleys below. I later learnt that this dam has an unusual convex design that adds to the impression of the water bulging against it.

Clywedog is the highest concrete dam in the UK, started after an Act of Parliament in 1963. It was completed in 1967 to reduce winter flooding and to provide drinking water for the West Midlands. These are simple sentences to write but they both hide and convey many deep layers. As the poet R. S. Thomas writes[6]:

> There are places in Wales I don't go:
> Reservoirs that are the subconscious
> Of a people, troubled far down
> With gravestones, chapels, villages even;

Reservoirs are built at huge historical, cultural and ecological cost. For many people in Wales, the slogan '*Cofiwch Dryweryn*' ('Remember Tryweryn') acts as a shorthand, as I best understand it, for the losses caused by the building of Tryweryn and other reservoirs such as Clywedog, as well as other types of incursions by the English into Welsh land, language and culture. '*Cofiwch Dryweryn*' refers, in particular, to the loss the Tryweryn valley and its village of about seventy inhabitants, Capel Celyn, situated between Bala and Ffestiniog in the Penllyn district in north Wales. In the 1950s, the Liverpool Corporation set its sights on the valley as a suitable location for a reservoir to provide water to their city, and in 1957 sponsored a private member's bill in the UK Parliament to approve the building of the reservoir. They took no regard of the village as custodian of the Welsh language and poetry and also of its important role as a stronghold for the local musical tradition of *penillion* – counterpoint singing of verses to the music of the harp. Despite much opposition (with thirty-five out of thirty-six Welsh Members of Parliament voting against, and one abstaining), the bill was finally passed in 1962, and the dam was completed in 1965. Three-hundred and twenty-five hectares (803 acres) of the valley were flooded: local ecologies, human and other-than-human, lost forever. Peaceful protest as well as sabotage, when a bomb destroyed equipment in 1963, served to delay, but not halt, the project.

The building of Clywedog reservoir between 1963 and 1967 followed on from this turbulent moment in Welsh history. There was no village in the valley here to act as a focal point for resistance and several farms had already been abandoned. One farm did have a particular historical significance. It had been one of first places to host the circulating Sunday school movement, important for the development of high levels of Welsh literacy in the nineteenth century.

During my stay in mid Wales, I was fortunate to share an evening with John. Now in his seventies, he had begun working at the Clywedog dam in 1966 at age eighteen as a fitter's mate in the machinery maintenance section. He described how Afon Clywedog was first contained across the construction site within a 2.4-m- (8-ft-) diameter pipe. The water cascaded from the pipe back into the river downstream of the construction site. The dam was then built at right angles to the pipe, which was later blocked up and removed so that the river then flooded the valley behind the dam. In 1966, there was a sabotage event here, too. A telephone warning was made to ensure no people were hurt but a small explosion destroyed the head mast. John describes how he and his partner Eileen now enjoy visiting the reservoir, to watch the birds there, re-living in his mind the valley bottom, in all the different stages of construction, lying below the still water. He often reflects on having been involved in such a significant and long-lasting project at the very start of his working life. I was particularly struck by one of John's reminiscences. When salmon, which had swum so very many miles to return to their spawning ground, reached the point where the river fell from the newly installed pipe, they were unable, despite many efforts, to jump up into it. Even if they had been successful, their spawning ground had been destroyed.

This haunting image of jumping salmon frustrated by human hand has stayed with me, reminding me of the high costs for salmon and other creatures and plants of providing humans with water: costs often not acknowledged by those who are the beneficiaries. I speak as one of those beneficiaries, growing up as I did in the West Midlands where drinking water includes water from Hafren and Clywedog. This supported my growth. Some of it may even remain in the long-living cells in my brain and eyes. When James urged me to drink water near Hafren's source I was not aware of the connection – this entanglement of water past and water present. Severn Trent, one of the water companies of the English Midlands, encourages careful use of water but persists in calling it a precious resource – something there for human use. What happens when we think instead, as some cultures do, of water as a lively being in its own right with active wisdom to share – if we are open to listening? In her poem 'At The River Clarion' (2009) Mary Oliver explores the importance of giving time and patience to listening to the voices of the wider natural world. They cannot be heard in a mere hour or day. As she comments, it is as though 'selfhood has stuffed your ears'.

Quiet time and reflection can be ways to 'unstuff our ears' and listen to the wisdom that the wider natural world, including rivers, has to share with us. Poetry and music can support such reflective listening, too. Keen to explore this possibility as part of my journey with Hafren, I approached the work of poet and lyricist John Ceiriog Hughes who spent his later years at Llanidloes and the nearby village of Caersws on Hafren's banks. He is probably best known today for the most widely sung Welsh lyrics for the traditional Welsh song 'Ar Hyd y Nos' ('All Through the Night'). He was born in 1832 at Pen-y-Bryn Farm overlooking the village of Llanarmon Dyffryn Ceiriog in the Ceiriog valley, Denbighshire, north Wales. Having shown aptitude for books and poetry rather than farming, he went to Manchester in 1849 where he obtained a job as a clerk in the railways and became involved in literary circles. In 1870, seeking a return to Wales, he moved to Llanidloes, and then, in 1871, to Caersws, 8 miles (12.8 km) downstream, where, until his death in 1887, he was the manager of the newly opened line between the village and the Van Lead Mine.

During his time in Manchester, Ceiriog Hughes helped to establish a literary society with three other Welsh poets: Creuddynfab (William Williams), Robert Jones Derfel and Idris Fychan. All three were influential on his work. Creuddynfab encouraged Ceiriog to move away from heroic and biblical themes and to adopt instead a simple, natural and popular style, similar to that of Robert Burns, exploring pastoral life and love. Derfel encouraged him to value and share the language and history of Wales and to adopt a bardic name. It was thus that Ceiriog inserted the name of his valley of birth between his first and family names. It was during his early period in Manchester that he wrote one of his much-loved pastoral poems, 'Nant y Mynydd' ('The Mountain Stream'),[7] in which he both identifies with and longs for the sparkling stream, the freedom of the small birds and the fresh mountain air of the valley of his birth. The poem ends:

Mab y Mynydd ydwyf innau,
　　Oddi cartref yn gwneud cân,
Ond mae'nghalon yn y mynydd
　　Efo'r grug a'r adar mân.

I am a son of the Mountain,
　　Far from home, making song,
But my heart is in the mountains
　　With the heather and the small birds.

Fychan was a singer to the harp and collector of *penillion* and traditional Welsh airs (tunes) and Ceiriog, too, developed a lifelong interest in collecting these. An important source of airs was Edward Jones's 1784 *Musical and Poetical Relicks of the Welsh Bards*. Ceiriog wrote poetry to accompany these tunes, enabling them to be sung in many homes and in public in the Victorian period. To appreciate his work, one needs to understand him as a lyricist. As Fychan put it, he had a remarkable gift for getting the feel of an air and incorporating its spirit into words. The musical nature of his work can now be appreciated via the many current performances of his work on the internet. Ceiriog's poetry and songs were also widely appreciated and enjoyed during his lifetime. With the exception of the Bible, his poetry collection *Oriau'r Hwyr* (*Evening Hours*), published in 1860, was the most-bought volume in Wales during this period. Thirty thousand copies were sold between 1860 and 1872.[8] On a visit to Cardiff's central library, it felt a tremendous privilege to hold in my hand a slim edition of these poems. The wafer-thin pages of the book were closely typed in tiny font, squeezing as much as possible into this clearly well-loved volume protected by a well-worn, extra cover made from paper with a red-and-green paisley pattern. It would have fitted easily into a pocket or on a side table. It was also a joy to

Growing

hold a larger, red-covered musical edition of *The Songs of Wales* that Ceiriog published with Brinley Richards in 1873, with the text in both Welsh and English. I wanted to hear the tunes played, and to join those who have heard these traditional songs spring from these pages. Ceiriog was awarded the bardic crown, the prize given for a free verse poem, at the Great Llangollen Eisteddfod of 1858, for his love poem 'Myfanwy Fychan'. This crown, which can now be seen at St Fagans National Museum of History in Cardiff, is made of real birch sprigs, leaves and catkins, which have been silvered and woven together to make a slim headdress that retains the birch's delicate beauty. This echoes the custom in the Ceiriog valley in which a girl gives a birch crown to her loved one.

Ceiriog was not, however, the only poet living near the Pumlumon Massif to win a crown. At the 1865 National Eisteddfod, held that year in Aberystwyth, the surprise winner of the crown, beating both Ceiriog and the Monmouthshire poet Islwyn, was Sarah Jane Rees. She is better known by her bardic name Cranogwen, after her home village of Llangrannog on the coast 30 miles (48 km) south of Aberystwyth, the town that lies below the western edge of the Pumlumon plateau. Her winning poem 'Y Fodrwy Briodasol' ('The Wedding Ring') is a humorous and somewhat sarcastic response to the destiny of married women.

Cranogwen was a remarkable woman, ahead of her times and now commemorated by a full-sized bronze sculpture in Llangrannog unveiled in 2023. Eschewing dressmaking, the path chosen for her by her parents, she followed her master mariner father to sea. She travelled widely and also studied for the exams necessary to become a master mariner herself, a role she achieved by the age of twenty-one. She then returned to her home village to be appointed, despite her young age, the head teacher of the school at nearby Pontgarreg. She also taught maritime studies to young people in the area. When we visited the village to see the new statue of Cranogwen we met Hannah, who served us lunch with a happy smile in the café there. Hannah was one of the last pupils at the old village school at

Pontgarreg before it closed in 2012. She enthusiastically told us that she remembers, at the age of five, being rather afraid of Cranogwen: a stern-looking women in a long black dress with a wide skirt and white collar looking down on her from the heavily framed Victorian photograph hung in the school assembly hall. As Hannah grew older, however, she began to understand and admire Cranogwen's achievements as a woman trailblazer and local hero and was proud to have attended the school where Cranogwen had been headmistress.

After her win at the National Eisteddfod, Cranogwen became an established preacher and lecturer, in the United States as well as at home. *Caniadau Cranogwen*, a volume of forty of her poems, was published in 1870. In her poetry she explored many themes ranging from the position/plight of women to romantic love. She also wrote poems such as 'Fy Ngwlad' ('My Country') and 'Dyffryn Cranog' ('Cranog Valley') expressing and celebrating the beauty and peacefulness of Wales, its green mountains, the softness of its breeze, and her love for her home village with its waterfall, birds and stream. Her poems are still available today in Welsh and English translation.[9]

In 1879 she became the editor of *Y Frythones* (*The Ancient British Woman*), only the second journal devoted to the interests of women in Wales, and the first to be edited by a woman. She was an ardent campaigner for temperance and for the protection of women from domestic violence, founding in 1901 the South Wales Women's Temperance Union. Although unable to fulfil her dream of opening a women's refuge before her death in 1916, one was completed in 1922 and named Llety Cranogwen (Cranogwen's Shelter) in her honour. Cranogwen had two long-term partners, Fanny Rees, who died at the age of 21, and later Jane Thomas. Buried at Llangrannog, her well-kept memorial high up in the churchyard has a long view looking down over the village and valley she loved. When I saw it, I was especially moved by the fresh green, purple and white ribbons newly tied to the railings around her grave – the colours of the suffragette movement – purple for freedom and dignity, white for purity and green for hope: a fitting tribute.

Growing

Having learnt so much about Ceiriog (as well as his friends and 'rivals') and keen to experience Ceiriog's mountains, streams and small birds, Peter and I set off from Llanidloes, leaving Hafren for a short while to visit Ceiriog's birthplace at the village of Llanarmon Dyffryn Ceiriog. Initially taking a route north-eastwards, we travel along main roads, broadly following the same route as Hafren. Like Hafren, we pass through Caersws, where Ceiriog had spent the last seven years of his life, and stop at St Gwynog's Church in the hamlet of Llanwnog to see his grave. Ceiriog's memorial, surrounded by an ornate railing, stands out strikingly in this rural churchyard set in the soft-green, rolling hills of mid Wales. A large white cross, adorned with a carved wreath, sits on a high white plinth, on which the details of Ceiriog's life are inscribed in Gothic script. Below this is the *englyn* (a traditional form of short Welsh poem that uses rigid patterns of syllables, rhyme and half rhyme) that he wrote as his epitaph:

Carodd eirau cerddorol, carodd feirdd,
Carodd fyw'n naturiol;
Carodd gerdd yn angerddol,
Dyma ei lwch, a dim lô.

He loved musical works, he loved poets,
He loved to live naturally;
He loved a poem passionately,
He is dust, and nothing remains.

We pause to reflect, then continue to follow Hafren's course until Welshpool, where we leave her to travel north through high-banked, twisting country lanes, occasionally glimpsing views of sloping green fields, sheep and isolated farms. It feels increasingly like a maze, and we wonder if we will ever escape. But cresting a hill, an unexpected, wide view of the Berwyn Mountains and the Ceiriog valley opens out before us. The wind buffets the frothy hedge parsley on the verges and the clouds form fast-moving dark shadows on the fields. Dipping downwards we soon arrive at Llanarmon Dyffryn Ceiriog, where two sixteenth-century inns, The Hand and The West Arms, stand guard over the village square. These hostelries once provided food and a place to rest to drovers and their animals travelling the various routes that converged here to cross over the stream on their way to markets in Oswestry, Chirk and Wrexham. We park by The Hand and, like travellers before us, enjoy some refreshment in the warm sunshine before walking past the village gardens bursting with roses to the bridge across 'Nant y Mynydd' – Ceiriog's mountain stream. Looking down from the small stone bridge, we can see the water flowing swiftly over the rocky bed. There is no public access, a notice sternly declares, but the landlady of the inn said it would be ok to enter here. So, a little nervously, we climb over the three-bar gate and make our way across the field and sit on soft tufty grass that grows beside the lively, tinkling water. Small birds do indeed dart amongst the rushes here, adding their song to the stream's voice. We sit quietly, listening for a while, absorbing the beauty and peace. I read aloud Ceiriog's 'Nant y Mynydd' and feel a connection, reaching back over the years, to Ceiriog and his love for this special place.

Returning to the lane, we walk up the very steep hill to Pen-y-Bryn Farm. The views all around us are stunning: verdant hills, dotted here and there with trees and hedges, slope gently into the curve of the valley with the glinting stream at its centre. The calls of sheep and lambs combine with birdsong and the burr of a tractor we can see in a field nearby. Here is peace, beauty: the epitome of a pastoral scene in sharp contrast to the busy, polluted urban sprawl in which Ceiriog found himself on his move to Manchester, and in which many of us find ourselves today. Pastoralism can provoke feelings of nostalgia, a yearning for an unchanging utopian idyll, but it can, instead, be approached in a different way – as a challenge to care for and protect the wider natural world. The children of this valley have expressed this so effectively in a video, with versions in

Growing

Welsh and English, they have posted online about Ceiriog and the poem 'Nant y Mynydd'. It shows views of their valley combined with animations they have drawn of Ceiriog's journey and time in smoke-filled, crowded Manchester. The poem is sung energetically, accompanied by the picture and sounds of the stream. Their rendition is lively, joyful, with loud clog dancing, and the final verse sung bravely, boldly – a call to action.

Growth and change are integral to all ecosystems: our planet and all within it, including streams and rivers. All are constantly evolving, emerging, each part adapting, changing. This is the paradox inherent at the heart of sustainability: that for something to endure, it must be allowed to change rather than be set into some kind of fixed nostalgic vision. However, for endurance to be possible, the change needs to be such that it does not overwhelm and destroy the ecosystem. In his pastoral song/poem 'Alun Mabon', Ceiriog draws attention to this relationship between time, change and endurance. It tells of the birth, life and death of shepherd Alun Mabon, the natural 'riches of a simple life he lives', guided by nature's daily and yearly cycles, unlike the life of a city dweller who is ruled by Big Ben's 'far booming sound' that 'bids city workers rise'. The final verse is an epitaph to Alun Mabon:

Aros mae'r mynyddau mawr,
  Rhuo trostynt mae y gwynt;
Clywir eto gyda'r wawr,
  Gân bugeiliaid megis cynt.
Eto tyfa'r llygad dydd,
  Ogylch traed y graig a'r bryn;
Ond bugeiliaid newydd sydd
  Ar yr hen fynyddoedd hyn.

Ar arferion Cymru gynt,
  Newid ddaeth o rod i rod;
Mae cenhedlaeth wedi mynd,

Hafren

A chenhedlaeth wedi dod.
Wedi oes dymestlog hir,
 Alun Mabon mwy nid yw;
Ond mae'r heniaith yn y tir,
 A'r alawon hen yn fyw.

The mighty mountains still stand,
 Wind roars over them;
Heard again at dawn,
 Shepherds sing as of old;
And daisies still flower
 Around the rocks and hills;
But there are new shepherds
 On these old mountains.

As the years pass
 Old Welsh customs are passed on and change,
A generation has gone,
 And a new generation has come.
After a long and turbulent life,
 Alun Mabon is no more;
But the ancient language lives on in the land,
 And the old songs endure.

Ceiriog lived at a time when it was still generally believed in Westernised cultures that although some humans were making harmful incursions into the wider natural and cultural world, overall these could be absorbed. 'The world was deathless' to use a phrase from the ancient Greek myth *Antigone*.[10] However, we now live in an unprecedented era where technologies are overwhelming ecosystems and permanently harming, or even destroying, them. We cannot take their survival for granted.

With birdsong and mountain streams at the forefront of our minds we reluctantly leave Llanarmon Dyffryn Ceiriog, nestled in

Growing

its beautiful valley, and retrace our steps back to Hafren at Caersws. We are spending the night here, ready to visit Porth Farm the next day. We had heard about the farm from the Montgomeryshire Wildlife Trust that had recently visited its woodland to look at the nature conservation ideas being developed there.

The following morning, after a most enjoyable Welsh breakfast, we leave the main road that runs beside Hafren as she flows along the valley bottom and turn into a small lane that winds gently up the side of the valley. Sheep are grazing in the green fields around us and grey-tinged clouds scurry across the soft-blue sky. We soon spot a large 'Welcome to Porth Farm' sign and pull in. Ahead of us we can see polytunnels for the farm's 'pick your own strawberries' scheme and, rising above these, a field divided into sections. One is filled with bright sunflowers, and another with pink, blue and white wild flowers, all tossing their heads in the breeze. We meet Gai, the farmer here, and he tells us about how making a living from farming is increasingly challenging for economic and environmental reasons. He has, therefore, decided to focus on ways to try to increase profit from his own acres rather than by renting additional land for sheep grazing. He has diversified into activities such as the pick-your-own flowers and strawberries we can see today as well as pick-your-own vegetables, a woodland walk and maize maze, an autumn pumpkin patch and Christmas activities. These bring in an income stream and reduce reliance on high-intensity sheep grazing with the damage that this causes to local plants and wildlife. This diversification also brings people onto the farm, increasing their understanding of, and connection with, the work involved in producing food and caring for the environment. These aims are reflected in the tagline 'Growing Together' Gai uses to promote his business on social media. The changes Gai is making have not been without challenges. Increasingly, stormy and unpredictable weather is making all the farming undertaken, including these new projects, ever more difficult. For example, in this particular year, the strawberry tunnels have only produced 10 per cent of the predicted yield. He strikes us, however, as someone who is determined to persevere, keen to find the balance needed between change, growth and conservation to ensure that the ecosystems and ways of life upon which farming depends can endure here at Porth.

Encouraged by our conversation with Gai, we set off to find the woodland walk he has opened up to visitors. Crossing a small bridge, we enter the wooded gully that a stream has formed in the hillside. We are free to find our own path and we set off towards the steepest section, thinking this will lead us to the top of the hill. The hawthorns, dotted with their developing deep-red berries, are too dense here for us to get through so we turn towards the more gently sloping section. The sun filters through a mix of oak, hazel, rowan and birch, dappling the woodland floor and highlighting acorns, leaves and the tiny pink wild geraniums nestling amongst them. Willows are growing towards the right-hand side of the wood near the gully where the stream runs in periods of higher rainfall. We hear the repetitive 'chiffchaff, chiffchaff' of the bird of the same name, the loud trill of a wren, and a song I don't recognise. Using my bird-identification app, I learn it is a willow warbler. These dainty birds, known for their solemn song, weigh only 9 g (0.3 oz) and are similar in size to a blue tit. They have a yellowish-olive breast and long, brown wing feathers, which allow them to undertake long winter migrations to areas south of the Sahara Desert. They forage for insects as well as eating berries in the autumn. They like to live in scrub and open woodland, near water such as the habitat provided here on Porth Farm. They are not yet listed as endangered even though their numbers have declined by a shocking 44 per cent in the UK since the 1970s. I am delighted and thankful to hear them here today.

Reaching the top of the woods, we emerge into the blustery open fields with a beautiful view down towards Hafren, the village of Caersws and the hills beyond: a rich tapestry of bright grass-green hues, edged with the deeper shades of trees and hedges, and dotted with the white highlights of sheep. After our climb, we are happy

Growing

to sit down at the edge of the field beside a thick hawthorn hedge. I lie back to feel the sun on my face and find that being beneath the hedge, out of the wind, is very comfortable. Peter lies back, too, and we are cosy, protected by the hawthorn from the wind that tosses the treetops around us. As we watch the ever-changing patterns of the clouds scurrying across the sky, I am reminded of Laurie Lee's description in his book *As I Walked Out One Midsummer Morning* of sleeping under hedges on his walk to Spain. I had always thought that this sounded damp and unpleasant, but I have learnt something new today. We close our eyes and doze. I begin to listen to the birds and hear a particular song. There are linnets here. Described by the sixteenth-century Welsh poet Edmwnd Prys as 'fellow poets', these small brown finches have a twittering, melodious song. Just over 13 cm (5 in) in length, they have a forked tail and white wing patches throughout the year. During the springtime breeding season, the males have a distinctive red patch on their chests. Mainly seed-eaters, they forage on the ground and in bushes for arable weeds including thistle, dock, hawthorn and dandelion. They favour dense hedges such as hawthorn for nesting, and require 'song-posts' such as those provided by low trees and fence posts. They have been on the British Birds of Conservation Concern Red List since 1996. Their numbers are in decline worldwide, and in the UK their population fell by 57 per cent between 1970 and 2014. This alarming reduction has mainly been driven by changes in agriculture, such as the removal of hedges and the use of herbicides. It is testimony to the care extended by Gai to nature here at Porth Farm that I can listen to this amazing songster. Lying here under a hawthorn hedge listening to linnets, Peter by my side, my heart is filled with joy and gratitude. This is my happiness.

A few days earlier in Llanidloes, I had seen a slate someone had hung from the railings at the place where Hafren and Clywedog meet. On one side of the slate was written, 'Have you enough?'. On the other it said, 'Enough is plenty'.

Chapter 3:
Powering
Newtown, Powys

Growth, change and endurance absorb my thoughts as we travel along Hafren's banks from Porth Farm at Caersws to the market town of Newtown (Y Drenewydd). Sheep graze the surrounding gently rolling green hills, and I think of the shepherd Alun Mabon and the challenging life that still endures here for farmers like Gai. In Newtown, we meet Sorrell, who works in the council offices. She shares with us her family's experiences with Hafren and Mochdre Brook, a tributary that joins Hafren at the entrance to the town, emphasising how confluences are 'magical places'. I ask her why she feels this, and she pauses. With a catch in her voice, she tells us how finding this still-wild place where two waterways meet means so much to her family since they moved to Newtown from the countryside. 'We thought we had lost all that,' she says. She goes on to tell us how the river beach there is made and remade, from sediment when the currents of Hafren and Mochdre collide. The children make sculptures and seating from logs. These exist, and then suddenly, they are gone – swept away by the lively water. New driftwood, new rocks appear, the shape of the beach changes. It is a constantly shifting micro-world.

Her family and friends share stories about the risings and fallings of Hafren and Mochdre – including extreme moments of change. Sorrell recounts a story told to her by her neighbour, who was related to a young girl, Bernice Haynes. In June 1936 after heavy rain, Bernice raised the alarm when the rising water of Mochdre Brook, where it joins Hafren, swept away the railway bridge. A train due to cross the bridge was halted just in time: disaster was averted. Sorrell suggests we visit this 'magical place' ourselves. She prints out a map for us, then lovingly draws on trees to help us find our way. It is hard to put into words the love and care with which she does this, but it hangs in the air.

Later, map in hand, Peter and I find our way through a housing estate to a footpath between trees and a large playing field where children are being coached on multiple football pitches. We reach Hafren, and the trees continue along her bank. The intense sounds of rapidly flowing water, the sway and rustle of branches and leaves, the sounds of twigs snapping underfoot, all transport us away from the town. In contrast, occasional discarded sweet and crisp packets, cigarette ends and drinks cans, as well as the cries of the children playing nearby, remind us that we are still very close to urban life. We reach the little beach that Sorrell so passionately described. Hafren, boosted by the water from Clywedog reservoir, is flowing rapidly, small waves lapping against the beach and the willows growing on the islets that have formed in Hafren's central flow. Clumps of sedges dot the water, and I reach out to one to examine it more closely, drawing on the traditional identification rhyme:

> Sedges have edges
> And rushes are round
> And grasses have knees which bend down to the ground.

Yes, definitely sedges, although nearby I can also see grasses with their knobbly joints at intervals along their stems. In contrast to Hafren's strong flow, Mochdre Brook has been reduced to a small, muddy trickle by the severe summer drought. It is hard to imagine it as a raging torrent capable of washing away a railway bridge. Yet flooding and deep, fast-flowing, powerful water are increasingly common – here in the UK and throughout the world – signs of systems pushed

Powering

close to or beyond the point of endurance. As we crouch down on the river beach I think of Sorrell and her connection to this easily overlooked place and the link to wild forces it provides for her and her family. Today, several branches have been washed up beside a circle of stones that have been laid here to create a small fireplace surrounded by logs set out as seating: I wonder how long these will be here and what will come next in this unassuming-yet-special place.

We make our way back to the centre of Newtown. Peter and I clamber down Hafren's bank, using exposed tree roots to help us, and we make our way across the exposed rocky ledges to perch beside the water. We watch a toddler and his father pick their way over the uneven rocks: a challenging adventure. The child bends down to pick up a stone and throws it into the water where it makes a pleasing splosh before disappearing. The child contentedly repeats this simple action. His father tells us it is his favourite game. Water is flowing rapidly, its insistent bubbling combining with the birdsong. I wriggle forwards on the rocks and step into Hafren's fast-flowing water. I feel the power of her current pulling against my legs. 'Careful you don't fall,' calls Peter from the bank as I wobble and raise my arms to steady myself. The muscles in my legs adjust to the pressure of the icy water and I take a careful step forwards, mindful of the slippery, soft moss on the stones beneath my feet. The air smells fresh and clean around me. I hear the high tinkling notes of the water as it tumbles over the rocks and the booming as it echoes in hollowed-out shaded places under the banks: these watery sounds intermingling with the cries of children playing in the large park nearby. I think of the tributaries that have added to Hafren's power. Some, like Mochdre Brook, can be approached easily. Others can be hard to find, although we may catch distant glimpses of them, as on my afternoon searching for Afon Dulas and Afon Brochan accompanied by the friendly bullocks. I think of the tributaries that have contributed to my own growth – some are easy to identify, others are only fleetingly perceived or even hidden, but that is ok, too. Life does not reveal itself in a single move. Standing here, reflecting on these tributaries, I feel grateful for the physical gift of water, including water from this very river that I drank from the tap in the English Midlands.

Although I am in central Newtown on this warm summer's day, I feel connected to a vital, untamed force, a deeper time. Across history, Hafren has linked peoples and transported goods, messages and animals. Iron Age people travelled along her route in long boats hollowed out from a single tree-trunk, and coracles made of skins stretched over willow and hazel frames. Romans navigated their way along her length, and she has powered industrialisation. I clamber back to the bank where my patient husband is waiting for me, reaching out his hand to haul me up on-to a rocky ledge. I have brought with me a recording of a friend, Mark, playing John Ceiriog Hughes's adaptation of 'Codiad yr Hedydd' ('The Rising of the Lark') set to music in *The Songs of Wales*. I play the music with the sound of the water rushing behind it, strong and dominant. The piano notes spin upwards, fall back, then rise again, higher and higher each time. I think of the skylarks that James and I saw at Pumlumon on our journey to Hafren's source and about how much Hafren has grown now to become this fast-flowing river. I feel emotions I cannot put into coherent sentences: pleasure in the beauty of this moment, sadness at the state of the world, fears for the future. Water sustains us but can also overwhelm us with her power.

The next day we make our way through the park and cross Newtown's Long Bridge, which joins the town's centre to an area called Penygloddfa. Hafren is flowing many metres below us and it is very hard to imagine the water rising so high it can barely pass underneath the arches of the bridge, even though we have seen photos of such high water levels during winter storms. We are on our way to the Newtown Textile Museum to learn about the development of industry in the town and its relationship with Hafren. Turning into Commercial Street, we see a row of front doors set in a four-storey building. The lower two floors are whitewashed and have small windows. The bricks in the upper floors are exposed, and the

Powering

windows are much larger. We are enthusiastically greeted by Sally, the secretary of the Museum Trust, who is on duty today. Sally's family have lived in Newtown since the 1820s and were the owners of the local tannery. Her interest in local history was sparked when her father, along with a number of other Newtownians, set up a civic society. In the early 1960s, they managed to get Penygloddfa designated as a conservation zone after there had been a substantial push to demolish much of the area. This led to an interest in the museum building, and in 1962 he wrote to the St Fagans National Museum of History, proposing a small museum be established in an historic set of buildings to preserve and tell the story of an important era of industrial development in Newtown.

For centuries, cloth called flannel – a corruption in English of *gwlân* or *gwlanen*,[11] the Welsh word for wool – had been produced on farms. Wool was sheared, washed, dyed and 'carded' (a process that smooths the fibres and causes them to lie parallel to one another). The wool was then spun and woven into cloth. The final stage was 'fulling', beating and compressing the cloth to soften the fabric and produce the distinctive warm, 'fluffy' feel of flannel. As the trade grew in the sixteenth and seventeenth centuries, surplus cloth was traded through the powerful Shrewsbury Drapers Company, and Welsh woollen flannel became a nationally and internationally traded product. Farm-based production changed when machinery developed. In the seventeenth century, corn mills in the town, powered by Hafren, were converted for fulling. It was not until the late eighteenth century, however, that carding and spinning were mechanised and added to the fulling mills in Newtown. Weaving, nevertheless, was still done on handlooms on the farms, which meant much transporting of goods was required. It made sense to bring the weaving processes into the town. Buildings, such as the one that now houses the Textile Museum, were, therefore, constructed early in the nineteenth century. Living areas were located on the ground and first floors and handloom weaving workshops occupied the top two floors, illuminated by large windows. Living conditions in these back-to-back poorly ventilated cottages were crowded and unhealthy, with one ground-floor room and one upstairs bedroom per family.

The museum today is a pleasant, airy, whitewashed space with the back-to-back cottages knocked through to create a bigger ground-floor area, twice the size of the original cottages, and with doors front and back allowing a through-flow of air. We walk around the living areas and the larger workshops on the upper storeys where we can see the wooden hand-weaving looms set up with their complex warp and weft. I try to imagine the very different environment the weavers and their families that lived and worked here would have experienced: the pungent smells, the clamour of children, the incessant clatter of the looms, and the dust and fibres in the air. We stop to admire the flannel samples produced on looms such as these. We rub them between our fingers and marvel at how the choice of wool and variations in the weaving and fulling process can produce the softest fabrics, their texture far from the itchy wool vests of our childhoods. The flannel is plain or has simple patterns such as blue, red or green stripes, which the early looms were able to produce, unlike the complex patterns associated with double-sided Welsh tapestry blankets made later in other areas of Wales.

The population of Newtown grew rapidly in the nineteenth century, increasing from approximately 990 in 1800 to 4,550 in 1830. In a survey conducted by the Handloom Weavers Commission in 1838, we learn that there were seventy-five manufacturers in Newtown with a total of 726 looms, employing 672 weavers.[12] Trade directories of the period indicate that there were also ten wool-carding establishments and ten fullers. By the 1840s, Newtown was called 'the Leeds of Wales', known for the high quality of the flannel produced. In the 1860s, steam-powered looms were introduced, and the weaving/living spaces were replaced by larger mills. These included the Oversevern Mill, the Cambrian Mill, the Commercial Mill and the Craigfryn Mill. Use of child labour was widespread, with company doctors reporting shifts of up to thirty-six hours for children aged seven upwards, and frequent accidents. The dependence of steam-powered mills on coal, however, put Newtown at a disadvantage, since the coal had to be brought in from a considerable distance. Other mill towns closer to coalfields started to dominate the industry.

The town did, though, continue to produce flannel, demand boosted by the activities of Pryce Pryce-Jones, a nineteenth-century Newtown draper. Seeing the opportunities created by the newly established Royal Mail and the developing rail network, he set up the world's first mail-order catalogue and began to sell flannel produced in his Cambrian Mill nationally and internationally. In 1889, he had about 40,000 customers, expanding to 250,000 by the end of the century. For customers in England, he promised next-day delivery from the huge warehouse he built next to the railway station, which even had its own post office. This building, now offices, still dominates the town's skyline today. Famous customers included Florence Nightingale and even Queen Victoria, whose bloomers were reputedly made from Newtown flannel. Pryce-Jones patented the Euklisia rug, a folding blanket with fasteners and built-in pillow – a forerunner of the modern sleeping bag. In 1887, he was knighted by Queen Victoria. In the end, however, the high costs of bringing in coal and the lack of direct rail links to the growing numbers of customers in the south Wales industrial valleys made flannel production in Newtown uncompetitive compared to the mills in northern England. The fire that destroyed the Cambrian Mill in 1912 effectively ended flannel production in Newtown.

This was not, however, the end of the textile industry in the Newtown area. Laura Ashley, famous for her British fashion and homeware brand, developed her business here. Laura and her husband publicly acknowledged the huge contribution of these Welsh towns and villages and their residents to the success of their international business. Born in 1925 at her grandmother's house in Dowlais near Merthyr Tydfil in south Wales, Laura attended school and the Hebron Baptist Chapel there. After a move to Croydon, she was then evacuated during the Second World War back to Wales where she attended secretarial college before returning to London. Whilst working as a secretary for the Women's Institute in London on a joint project with the Victoria and Albert Museum on women's handicrafts, Laura realised that there were not many fabrics available using traditional Victorian designs. She, therefore, began to use details from original prints to create her own designs for headscarves and tea-towels. Her husband Bernard, an engineer by training, although by then working in the City of London, printed these in the kitchen of their small London flat using a small hand-printing frame he had designed. Thus began a global business. Demand took off and they opened a small production unit in Kent but, unable to find a place to expand, looked to Wales, relocating to Machynlleth in 1961.

In 1967, they moved to Carno, a village nestled in the hills 12 miles (19.3 km) north of Newtown, where they developed their first factory. Laura Ashley was keen to support the local women and families working for her (she had four children herself). Women with young children were allowed to collect their children from the local primary school at 3.30 p.m., and Laura herself collected her youngest daughter, Emma. They could also leave work at midday on Fridays to do their family shopping. Healthy lunches were provided each day in the canteen and there were women's football competitions at lunchtime. Homeworking was also possible. In 1984, a modernised company headquarters was opened in Newtown creating 500 jobs, and around this time, approximately 750 people were employed by Laura Ashley and its subsidiary, Texplan Manufacturing, in mid Wales. At the height of the brand's popularity, there were over 5,000 Laura Ashley retail outlets around the world, including in Paris, Geneva and Tokyo. Once again, products designed and manufactured in Newtown, as well as in factories in Wrexham and Machynlleth, were being sold worldwide.

On the visit Peter and I had made the previous year to Machynlleth, on our way to Glaslyn Nature Reserve on Pumlumon, we had met

Jean. Happily reminiscing about her time at the Laura Ashley Texplan factory in Machynlleth, Jean told us how she used to sew full skirts and square-necked tops. From her voice and expression, we could tell it clearly meant a lot to her to have been part of this famous fashion story. When I said how lucky we felt to meet someone who had worked for the company, she replied: 'The Laura Ashley business was so big around here that pretty much every local of a certain age you speak to would either have worked there or known someone who did.' We shared our teenage memories, the excitement of going to the Laura Ashley shops in Oxford and Chester, how sophisticated it made us feel. Jean added happily, 'I was a part-time hippy then – in the evenings and weekends. I had to wear my school uniform the rest of the time!'

In 1985, when Laura Ashley died, aged just 60, after a fall on the stairs, the company was on the brink of expansion. It was floated on the stock exchange with a valuation of £200 million in 1986. However, changing tastes and the loss of Laura Ashley's creative input led to its steady decline. In 2020, as the global Covid-19 pandemic took hold, the business finally went into receivership and the 77 employees in the remaining small Newtown manufacturing and distribution centre lost their jobs. This is not, however, the end of this particular textile story. The social enterprise Into Fashion decided to invest, with the support of the Welsh Government. They described the skills of the workforce, many of them having over thirty years of experience, as 'gold dust'. Within a month of opening, they were producing 20,000 garments per week.

Reflecting on Newtown's textile histories, old and new, we leave the Textile Museum and return to the Long Bridge. From here, we can see down Broad Street, where the busy weekly market is trading. Today, Newtown is an attractive small town with some light industry and tourism. At the end of Broad Street, sheep graze on the green hills that rise steeply behind the clocktower of the Town Hall. This houses, on its ground floor, a museum dedicated to Robert Owen, widely heralded as the father of the co-operative movement and a pioneer of socialism. Owen was born in Newtown in 1771. He became interested in the textile trade whilst working in a neighbour's drapery shop. He left home aged just ten to become a draper's apprentice, then moved to Manchester in 1787. By the age of twenty-one he was a manager of one of Manchester's largest textile mills, becoming, at just twenty-three, a managing partner at the Chorlton Twist Company.

Whilst travelling in Scotland he met his wife Caroline Dale, and in 1799, with some like-minded business partners, he bought his father-in-law's mill at New Lanark. Here, he introduced a very different approach to manufacturing and the treatment of workers, an outlook that ensured he made his mark on history. Contrary to the thinking of the day, Owen believed that a person's character was formed by their environment rather than natural inclination. He believed that with better food, housing, working conditions and education a person could be healthy and happy, and consequently, drunkenness and crime would reduce: radical ideas at that time. In New Lanark he wanted to create a community where the interests of owners and workers were not in conflict. In an era when children as young as five were often working long hours in dangerous conditions, he raised the starting-age for child workers to ten and established a nursery and schools for the younger children. He improved housing, reduced working hours for all and introduced a sickness insurance scheme as well as establishing evening schools and recreation halls for adult workers. In the mill he introduced a 'silent monitor': a block with different coloured sides that was turned to display different levels of output. Good output received extra reward. Under Owen's skilled management, the mill was financially successful, too, although his business and personal ideas, including his resistance to organised religion, caused conflict with some of his partners. The New Lanark Mill is now preserved as a UNESCO World Heritage Site.

In 1824, wanting to take his social experiments further, Owen used his own capital to set up the New Harmony settlement in the United States. His vision was for a self-sufficient co-operative communal village that would act as a beacon of utopian socialism and encourage others to set up similar villages until they were widespread through society. Unfortunately, the village itself failed due to a lack of practical skills and the frictions of communal living, although it did successfully sow seeds for future social and political change. In 1828, Owen returned to London with the majority of his fortune lost, but nevertheless, he was undeterred. He continued to campaign for workers, and his activities included setting up the short-lived Grand National Consolidated Trades' Union in 1834. Despite the brief life of this organisation, due to concerted opposition from employers, the government and the courts, the Union was nevertheless influential in the development of the future trade union movement. Owen continued to campaign for co-operative principles, factory reform and education for working-class children. His followers included the Rochdale Pioneers who established the first co-operative store in 1844. Knowing his end was near and wanting to be buried alongside his parents, Owen returned to Newtown in 1844. The landlord of The Bear Inn gave him a room free of charge and it was there that he died at the age of 87. He spent his long life and a large fortune introducing ideas that, whilst not always successful in his lifetime, acted as jumping-off points for social change. On his deathbed he said: 'My life was not useless. I gave important truths to the world, and it was only for want of understanding they were disregarded. I have been ahead of my time.'

I am very moved by all we have learnt at the museum. Owen was so forward-thinking and concerned for his fellow men, women and children. He never gave up, despite setbacks, but worked until the end, campaigning and experimenting with new ideas. This is an important lesson in a world where new ways of thinking and being are still needed. I feel the challenge he sets us. We take the short walk northwards from the museum to Owen's grave at the ruined medieval church of St Mary's, built on Hafren's banks but abandoned in 1856 due to frequent flooding. Owen's tomb, inscribed with his name and those of his parents, Robert and Anne, sits against the southern wall, away from the riverbank. The tomb is enclosed in decorative railings in an art nouveau style – a later addition dedicated in 1902 to honour his contribution to co-operative and socialist movements. The railings feature a plaque by Alfred Toft, on which Owen is surrounded by labourers and the slogan 'Each for All', a quote from the co-operative movement's motto 'Each for All and All for Each'. We sit on a sunny bench and reflect on Owen's ideas, his contributions to education and factory reform, and the continuing need to address exploitation in world trade today. Are the red flowers around us a deliberate choice, we wonder? The bees are certainly enjoying their presence.

We walk around the church to look over the wall that borders the graveyard. We expect to see the riverbank close by, given the church's history of flooding. Instead, we see Hafren a long way below. Puzzled, we recall Sally at the Textile Museum telling us about the flood defences built in Newtown in the 1960s and decide to find out more. Newtown is built within an upside-down U-shaped loop in Hafren's course. A natural moat was an attractive proposition when, in 1279, Edward I of England granted the right to his powerful ally, Roger de Mortimer, to build a new town on territory seized from the Welsh. However, the flooding this creates had always been a threat to the town's survival, especially as it expanded closer to Hafren's banks, and storms had become more severe and frequent. In 1960, the town experienced severe flooding. When this happened again in 1965, many felt that the future of the town was doomed; homes and businesses could no longer get insurance and the town's council could not afford to build flood defences. However, in 1967 Newtown was granted 'new town' status by the UK Government. This released funding for flood defences with a focus on hard engineering projects. High embankments were constructed to the east of Long Bridge as this was an area where water frequently overspilt its banks and rushed

Powering

across the town. High flood walls were built in other areas. On the eastern side, a new route for the river was constructed to remove a small additional loop in its course that slowed the exit of water from the town. A rare black poplar tree, previously on the north bank of the river, now finds itself on the southern one – what a phenomenon: a 'walking tree'.

Combined with the control of water flow at the Clywedog dam, the new flood defences were effective. But more recent increasing winter storms now mean that flooding is once again a problem. Focus has now shifted towards 'soft engineering' that aims to reduce the flow and speed of water into the town, alongside acceptance that green spaces should be left for floodwater. I had learnt from Sorrell at the council offices of an exciting new project to work *with* Hafren, to enjoy her and see her as valuable rather than as a 'liability'. Since 2015, the council, which covers Newtown and the nearby village of Llanllwchaiarn, has been working on a project with Open Newtown,[13] a consortium of local volunteer groups, community enterprises and organisations. The aim of the project is to protect and manage Newtown's green and blue spaces more imaginatively and sustainably with facilities and activities for locals and visitors alike. In 2018, the consortium was awarded £1.1 million of Big Lottery Funding to support the development of riverside meeting spaces, such as Hafan yr Afon ('Haven by the River'). This space opened in 2022 and provides meeting rooms and a hub for various activities including nature adventure, forest and river schools and access to the water.

Access to Hafren is an important issue that takes me by surprise when I first visit her upper reaches. As a child, I had washed my dolls and later built dams in streams that find their way to Hafren. As an adult, I often walk on footpaths along Hafren's banks in Gloucestershire and Worcestershire, and I am shocked to discover this is not always easy all along her course. Naively, I hadn't understood that rivers are 'owned' up to their midline by the owner of that portion of the bank, and this confers fishing rights that can then be sold to angling societies. There is no missing this point, however, when we decide to take an evening stroll in the countryside around Newtown. In a single layby beside a bridge over Hafren, just east of Newtown, we see the following nine signs:

> PRIVATE FISHING
> MEMBERS ONLY
> NO NIGHT FISHING
> NO DOGS
> WARNING: OPERATION LEVIATHAN
> NO DAY PERMITS
> WARNING: THESE PREMISES ARE PROTECTED BY
> CLOSED CIRCUIT TV
> NO PARKING
> NO DOGS ALLOWED

Excluded, I stand on the bridge looking at the rushing waters below. Access gates are padlocked, and a single fisherman stands on the banks. I feel both sad and angry at my exclusion and at the reduction of Hafren to a commodity to be traded.

These feelings of anger resurface the next day when I visit the weir over Hafren at the nature reserve at Llanllwchaiarn, a mile downstream from Newtown. I want to see the place where John, who was involved in the building of Clywedog reservoir, goes to sit quietly in the hope of seeing salmon leaping. The footpath to the weir is along the filled-in Montgomery Canal, rather than beside Hafren. Spotting a gap in the wire fencing, I slip through to a small grass expanse that leads down to the riverbank. Reaching Hafren, I kneel down on the soft, damp grass and lean forwards to trail my fingers in the cold water. The traffic noise from the A483 is strong here but so, too, are Hafren's soothing watery notes. Green is the dominant colour: the grassy banks are fringed with trees that are dipping their leaves into the grey-green water. Reluctantly, I walk on and reach the nature reserve, which is tucked in behind Newtown's sewage works. A passer-by walking their dog promises me that the

Powering

footpath eventually leads to a wide area beside Hafren, and before long, I can hear the chatter of teenagers. The footpath opens up, and I can see the weir. To the side of this is a 'ladder' to assist salmon journeying upstream, although John had told me he has never seen salmon using this as they seem to prefer jumping up over the weir. The water is deep and dark, almost black. It is an uninviting spot, tucked in behind the sewage works. The youngsters are piled on top of each other on a single bench. They are happy to chat with me and tell me a rather jumbled story of one of them having to go into the water to rescue their shoe. This feels a dangerous place and I learn later from John that the bench the youngsters were sitting on is a memorial to a young canoeist who drowned at the weir. I feel anger for these young people. Is this the best access that can be provided for them? There is perhaps a stage in one's teenage years when one is drawn to 'darker', lonely spots, without adult supervision or structure. But surely something safer could and should be provided for these and other young people along Hafren's route?

In the giant's story, Hafren tells us how she will be kind to the people on her route, bringing them water. However, if they abuse their position, she will rise up and show them her power. In Newtown, we have seen Hafren's power. She helped to drive the textile industry that flourished here, and now increasingly provides a focus for leisure and tourism. She also shows her power when rising up and using her waters to warn us of the impacts of climate change and human abuse of the wider natural world. We need to find ways to live with water, ones that look beyond containing it and instead focus on learning to live with and alongside it. Power has also played its hand in other ways in Newtown. Standing here beside Hafren's fast-flowing water at Llanllwchaiarn, I question the power that excludes the majority so determinedly from her banks along many stretches nearby and that brings these young people to this accessible but dark and dangerous spot. I accept it is important to prevent overfishing and protect Hafren's banks from damage, but surely this has to be balanced with the need for access to all she freely offers?

Chapter 4:
Bordering and crossing
from Rhyd Chwima, Powys, to Shrewsbury, Shropshire

Peter and I sit side by side on a seat perfectly formed from the exposed roots of a willow. Grey clouds and weak sunshine are reflected in Hafren's silvery, fast-flowing water before us, their shapes broken up, distorted. We have travelled 4 miles (6.4 km) downstream from Newtown and are now at Rhyd Chwima (Swift Ford), near Montgomery, grateful to have found such an accessible spot to enjoy. Around us, willows dip their branches into the dappled water and the opposite bank is a riot of invasive pink-flowering Himalayan knotweed mixed in amongst the tall, waving fronds of

Bordering and crossing

grasses and sedges. We are happy to take time to rest here, to listen to the birdsong mingling with the gurgling of the water and watch the bubbles that form as it tinkles over the rocky riverbed. I step into the water but its chill, strong current and depth, boosted by the water released from Clywedog reservoir in these drier summer months, quickly deter me from trying to cross. I imagine the sights and sounds of the many people who have stepped into this water before me, or ridden across, the water splashing upwards from the horses' hooves. The peace here today belies this place's history as a strategic point to cross the boundary created by Hafren between different territories and kingdoms.

The ford here, known as Allwedd i'r Deyrnas (The Key to the Kingdom), has been well defended throughout the ages. A sign beside the footpath informs us of the well-preserved ramparts of Ffridd Faldwyn, the Iron Age hill fort a mile south-east of the ford, and the remains of Forden Gaer, a typical Roman fortified camp in the field north of the ford. Thought to be Lavobrinta, it was occupied until 300 CE and was one of the two largest camps in Wales. Offa's Dyke, the extensive fortifications built in the 780s by Offa, King of Mercia, to defend his borders, runs 1.5 miles (2.4 km) east of the ford. In the eleventh century, the Norman Roger de Montgomery built a motte-and-bailey castle at Hen Domen and the remains of this can still be seen 0.5 miles (0.8 km) south-east of the ford. One mile (1.6 km) north-west of the ford is Montgomery Castle, established in 1223 during the reign of Henry III. Montgomery Castle played an important role as a place of negotiation in the military campaigns between Henry and Llywelyn ap Gruffudd, and it was here at the ford that representatives of Henry and Llywelyn signed the Treaty of Montgomery on the 29 September 1267. This treaty ceded considerable autonomy to Wales and recognised Llywelyn as the Prince of Wales – an autonomy that was, however, short-lived. It was crushed by Henry's son Edward I who declared that henceforth it would be the firstborn sons of English kings who would hold the title 'Prince of Wales'.

Hafren

Hafren has long acted as a border between England and Wales. During the Roman settlement of Britain, the military leader Plautius was only able to subdue the British in an area bounded by Hafren and the Trent before returning to Rome in 47 CE. Continuing Roman campaigns did eventually subdue resistance from local tribes such as the Silures. In 70 CE, the Roman leader Scapula crossed Hafren at Caersws, (*caer* is Welsh for 'fort' or 'stronghold') and this became an important Roman defensive settlement. An initial campaign fort was replaced by a larger settlement and the earthworks of this remain clearly visible today. John Ceiriog Hughes would have seen these when he lived at Caersws towards the end of his life. The railway station there is built within the ramparts of the fort. During the Saxon invasion of Britain, Hafren once again marked an important border: in 577 CE, at the battle of Deorham (Dyrham) in Gloucestershire, the Saxon prince Ceawlin defeated the Celtic Britons. Some retreated across Hafren and were thus cut off from their fellow Celts who, also defeated by the Saxons, retreated into England's south-west peninsula, today's Cornwall.

The connection between these Celtic Britons is widely recorded in myth and legend such as the tales of Arthur's court in the *Mabinogion*. This famous medieval Welsh text records eleven stories from the long-standing Welsh oral tradition, dating back many centuries, combining Celtic mythology and Arthurian romance. Preserved in written form in the *White Book of Rhydderch* (1300–25) and the *Red Book of Hergest* (1375–1425), fragments of texts have survived from earlier periods. Exact dating of the stories is problematic, but portions of the stories are believed to date from as early as the second half of the eleventh century, with some stories believed to be much older still. The stories became popular in the nineteenth century when Lady Charlotte Guest translated and published a collection of eleven Welsh tales using the title the *Mabinogion*. This title is actually a misnomer, as Lady Charlotte mistakenly believed that *Mabinogion* was the plural of *Mabinogi*, a term derived from the Welsh word *mab*

Bordering and crossing

(boyhood/youth), which then developed a meaning of 'tale of a hero's boyhood' and eventually, simply, 'a tale'. Many editions of the stories now exist both in Welsh and in English translation and the wide availability of audiobooks makes it easy – once again – to enjoy these myths in their original spoken form. In the story 'Culhwch and Olwen', believed by many to be one of the oldest in the collection, Culhwch travels from Wales to Cornwall to seek support from his cousin King Arthur in his quest to marry Olwen, the daughter of Ysbaddaden of Bencawr, a powerful and ruthless king of the giants. Culhwch asks Arthur to cut his hair, and Arthur performs this ritual before his court as a sign that he acknowledges the kinship between the two of them. In the legend, Culhwch must undertake complex quests, perform perilous feats and retrieve seemingly impossible-to-obtain treasures to win Olwen's hand. Arthur provides men to Culhwch, and sometimes even gets involved himself. Many adventures and battles ensue, including sailing to Brittany to find the two dogs of Glythfyr and to Ireland to ask for the Cauldron of Diwrnach. Hafren's banks provide both a place to meet and a place for Arthur to ambush his enemy, Twrch Trwyth, and drive him into Hafren's waters. Pumlumon is mentioned as a vantage point in the search for Dillus Farfog, whose beard is needed to create the only leash strong enough to hold Drudwyn, the whelp of Greid, son of Eri. After many twists and turns, Culhwch does eventually win Olwen's hand.

Traditional Gloucestershire stories[14] also tell of how Ceawlin and his Saxon forces engaged with the retreating Celtic Britons. One story tells us of an action near Fretherne, a settlement on the Arlingham peninsula near Gloucester. This peninsula, created by an extreme meander in Hafren's course, has a striking, almost eerie atmosphere to this day, with Hafren appearing at every turn. On seeing the Celtic Britons fleeing across Hafren at Priddy's Point, the Saxons plunged into the water at nearby Unla Point, believing they could cut off their retreat. Unknown to these invading Saxons, this is one of Hafren's most dangerous stretches. She is

Hafren

tidal here, but even at low tide crossing is treacherous with many currents and whirlpools in the remaining channels of water. Many drowned, and the word 'unla' is thought to be a contraction of the Saxon word for misfortune. Saxon presence is still very visible all along Hafren's left (eastern) banks in Gloucestershire. In nearby Deerhurst, there is still a Saxon font in St Mary's Church, and the Saxon Odda's Chapel was discovered within the timbers of a farmhouse there. Hafren thus played a major role as a boundary, helping to preserve Celtic traditions during this period of history. The name 'Wales' comes from the Saxon word *wealas*, for 'foreigner', 'stranger' – a people unknown and unconquered. In contrast, the root of the word 'Cymru', the name used in Welsh, is 'fellow countrymen' and was already seen in early Welsh texts such as the *Book of Taliesin*, a famous fourteenth-century manuscript written in Middle Welsh (a form of Welsh that developed in the twelfth to fifteenth centuries). Many of the fifty-six poems it records are believed to date back to the oral traditions of the tenth century or earlier, possibly even to a sixth-century real poet (or group of poets) known as Taliesin.

Following the Norman conquest of Anglo-Saxon England in 1066, William I installed three of his most trusted lords, Hugh d'Avranches, Roger de Montgomery and William FitzOsbern, as earls of Chester, Shrewsbury and Hereford – strategic points along the borders with Wales – establishing an area known as the Welsh Marches. The term is derived from the Anglo-Saxon word *mearc* that means 'border' or 'frontier'. The marcher earls had 'petty kingdoms', which were largely independent from the king. Over the next four centuries, the earls established more, smaller, marcher lordships to control these borderlands. The area had the largest number of motte-and-bailey castles in England and Wales. However, the independent power of the marcher lords began to wane as more of the lordships were returned to the crown and independence was finally ended by Henry VIII in the sixteenth century. This borderland country, particularly Shropshire, Hereford and parts of

Bordering and crossing

Powys, Monmouthshire and Wrexham, is still sometimes referred to as the Welsh Marches, its historical role as an in-between, contested borderland lingering into the present day.

Hafren also appears as a battle site in the opening act of Shakespeare's *Henry IV, Part 1*. Shakespeare (somewhat inaccurately) depicts the battle between Owain Glyndŵr and Edmund de Mortimer on 'gentle Severn's sedgy bank in single opposition hand to hand'. Owain Glyndŵr, a descendant of several of Wales's royal dynasties, led a fifteen-year rebellion, commencing in 1400, against English rule in Wales. Over a number of years, he gained control of several key castles, including Aberystwyth and Harlech, and attracted international support, for example, from the French. He was crowned Prince of Wales and, in 1404, held his first Senedd (Parliament) in Machynlleth, setting out his plans for the country. His military success, however, began to wane, and in 1409, after a long siege, Harlech castle fell to the English. Glyndŵr managed to escape and led several more skirmishes against the English. Despite the offer of large rewards for information, he was never recaptured. In 1413, Henry V came to the English throne, and with a more reconciliatory approach to the Welsh, offered two pardons to Glyndŵr that were never accepted. No official record of Glyndŵr's death exists. It is believed he died in 1415 although no known grave exists. He has obtained an almost mythical status, with tales saying he/his spirit lives on in the mountain caves of mid Wales. A long-distance footpath, named for Owain Glyndŵr and the routes he trod, passes through the upland bogs of Pumlumon. Peter and I had crossed this on our walk at Glaslyn lake and nature reserve the previous year. The footpath headed upwards to a lonely crag, a buzzard circling overhead. A lonely spot indeed: it was not hard to imagine a legendary presence hiding out there.

As so often is the case in recorded history, many stories of Hafren's role as a border have a strong militaristic focus. However, her role as a border in everyday life was also significant. Places to cross – fords and, later, bridges – stand as testimony to this, as do the inns, villages and towns on her banks. Sitting beside Hafren here at Rhyd Chwima in Powys, I think back to a visit Peter and I had made to the Old Passage Inn on the Arlingham peninsula in Gloucestershire. This stands today as testimony to the ferry and ford that existed there for many centuries: a safe crossing-point unlike the treacherous water at nearby Unla Point where the Saxon warriors perished. We had visited the inn on a scorching summer's day, approaching it along a flat, narrow road. No longer a place of crossing, we could only gaze across to the opposite bank and the village of Newnham. The mud banks, revealed at low tide, glistened in the sunshine, and the cerulean-blue sky was reflected in the thin ribbons of water that still remained. There were no trees offering shade as we walked along the public footpath that ran alongside Hafren. I thought of the teams of men who had pulled the trows, flat-bottom cargo boats built for Hafren's shallow waters, back up-river along these paths, and the demands and harshness of their lives. Like these travellers of the past, we were glad to return to the inn overlooking the water and the shade and refreshment it offered – for us in the form of ice cream that we happily devoured in the garden there. No doubt the trowmen preferred something a little stronger.

Hafren no longer forms the border in Gloucestershire between England and Wales. This has shifted further west and is now unmarked, crossable with ease, a situation we have come to take for granted. The border did, however, take on new significance during the Covid-19 pandemic. Devolution of power, which had begun in 1998, with the UK Parliament's Government of Wales Act, meant that Wales, now with its own Senedd (Parliament),[15] could issue different Covid-19 public health laws and travel restrictions from those issued in England. On many occasions during those traumatic months, we gazed from our vantage point in South Gloucestershire across Hafren's estuary towards the Black Mountains in Wales, unable to enter Wales or see my sister in Cardiff.

Rivers have long histories of acting as physical boundaries. They also exist in our imaginaries as metaphysical ones: liminal, transitional

Bordering and crossing

spaces between 'here and there' where danger as well as change and rebirth can occur and entrance into other worlds is possible. In Hafren's tale, when the young princess is forced into the river by her stepmother, she is taken into the care of the river nymphs and reborn as the river's goddess. In early Christian baptisms, submergence in the river Jordan washed away the sins of the old life and generated rebirth into a new one: a process symbolised in Christian baptism to this day by the application of holy water. Rivers, such as the river Ganges for Hindus, have significance in many religions. The positioning of such otherworlds is also significant. In Greek mythology, one entered the underworld – a separate realm under the earth (or in some stories situated at the periphery of the world) – by being ferried across the River Styx. In the great Welsh Christian hymn 'Guide Me, O Thou Great Jehovah'[16] by William Williams, the singer asks that when he 'treads the verge of Jordan' Jehovah will land 'him safe on Canaan's side'. Heaven and hell are separate realms outside and beyond worldly existence.

In Celtic myth and legend, however, Annwn – an 'otherworld' of pleasure and plenty – exists alongside this world. One can enter it through liminal places such as rivers and cauldrons or by being led there by mythical creatures. There is a 'to-ing and fro-ing' between these realms that are full of energy, liveliness and enchantment. Boundaries of many kinds shift, collapse and are remade, as the *Mabinogion* vividly portrays. In the First Branch (section), Pwyll, Prince of Dyfed, is out hunting in Glyn Culch when he is separated from his party. He comes across the mythical creatures Cŵn Annwn, the dazzlingly bright spectral-hounds of Arawn, King of Annwn. The hounds are feasting on a fallen stag and, in a breach of hunting convention and not knowing whose dogs they are, he drives the creatures away so that his own dogs can eat. King Arawn arrives and takes Pwyll to task for this discourtesy. Pwyll offers to makes amends. Arawn asks to exchange places with him for a year and a day, and then on the last day for Pwyll to kill Arawn's rival King Hafgan. Pwyll takes on Arawn's appearance and spends a year enjoying court life and plentiful hunting, although he does not take advantage of the exchange to have relations with Arawn's wife. As promised, on the last day he defeats Hafgan and returns to Dyfed. A strong friendship begins between the two kings even though they live in different realms. They share gifts of greyhounds, hawks and other kinds of treasures. Pwyll became so prosperous that he became known as Pwyll Pen Annwn (Pwyll, Lord of Annwn/the Otherworld).

Celtic myths also invite us to explore the boundaries between human and the other-than-human that can be crossed, remade, changed. One of my favourite stories is the tale of Lleu Llaw Gyffes in the Fourth Branch of the *Mabinogion*. Lleu is unable to marry a human bride due to a curse placed on him by his mother, so his uncle, Gwydion, and Math, King of Gwynedd, fashion the most beautiful bride for him out of flowers of broom, meadowsweet and oak. They name her Blodeuwedd, which means 'flower face'. I like the idea of being fashioned from flowers! However, Blodeuwedd does not turn out to be fair of nature. She has an affair with Gronw Pebr, Lord of Penllyn, and they plot Lleu's murder: a difficult task since Lleu can only be killed by the solving of a seemingly impossible puzzle. Blodeuwedd tricks Lleu into revealing that he could only be killed at dusk, by a spear that had been forged for a year, during the hours when everyone is at mass. Lleu must have one foot on a bath tub set on a river bank under a thatched shelter and the other foot on a billy goat. She conspires to place him in this exact position whereby Gronw throws the correctly forged spear. Wounded, Lleu turns into an eagle and flies away, and Gronw and Blodeuwedd rule Lleu's lands. However, Lleu is tracked down by his uncle, Gwydion, who lures him down from the tree where he is sheltering by singing an *englyn* (a traditional Welsh poem). Gwydion nurses Lleu back to health, and together, they defeat the usurpers, killing Gronw and turning Blodeuwedd into an owl. Her eternal punishment is not to see daylight, to be attacked by other birds and to live a lonely, solitary life.

Bordering and crossing

Shapeshifting is also a key part of the legendary story[17] of the birth of Taliesin, the sixth-century Welsh poet with mythical, as well as possibly historical, status. The sorceress Ceridwen is the keeper of the cauldron of poesy. She has two children: a beautiful daughter and a hideous son. To compensate for his looks, Ceridwen brews a magic potion that will make him all-knowing and infinitely wise. The potion needs to be stirred for a year and a day, so she hires a man named Gwion for this task. A year and one day pass and, whilst Gwion is stirring the potion, three drops splatter onto his finger. He licks his finger to soothe the burn. He thus becomes the unwitting recipient of the powers of the potion. The cauldron bursts open and the remainder of the potion is lost. Furious at this turn of events, Ceridwen chases Gwion and they shapeshift into various animals – hunter and prey – as the chase progresses. Firstly, Ceridwen is a greyhound and Gwion a hare, then they become an otter chasing a fish, then a hawk pursuing a small bird. In the final shift, Ceridwen becomes a hen and Gwion a grain of corn, which she swallows. Nine months later, she gives birth to a baby; Gwion is reborn. Unable to kill such a beautiful baby, she places him in a coracle, and, as in of the story of Moses told in the Torah, the Bible and the Koran, she sets him in the river. He is found by Elphin, a courtier of the sixth-century King Maelgwn. The baby was named Taliesin, which means 'shining brow', and he grew up to be a leading satirist, poet and prophet of his day. Whilst it is discussed whether all the poems in the *Book of Taliesin* are attributable to him, these poems are undoubtedly a key collection of early Welsh stories and ideas.

Taliesin's mythical beginning sits well with a key aspect of a poet's gift – that of *awen* – poetic inspiration that goes beyond bardic skills of poetic forms and knowledge of ancient lore, geography and history, to a state of altered consciousness. In this state, a poet receives knowledge of matters beyond what can be routinely learnt: an inhabiting of and a shapeshifting into a life other than one's own. In 'The Battle of the Trees' Taliesin declares his wide-ranging experience of *awen*, from being 'a slim enchanted sword' to being 'a bridge for crossing', a bubble, a bee, an eagle and a 'raindrop in a shower'. *Awen* gives him poetic prowess and an advantage over his rivals.

The stories and poems of the *Mabinogion* and the *Book of Taliesin* are wonderful in their own right, and I highly recommend them. They also have important wisdom to share in this time of ecological crisis. They challenge the disenchanted age in which those in Westernised society live. As the ecofeminist Val Plumwood put it, Westernised humans hyper-separate themselves from the wider natural world, first, by seeing themselves as a separate species, and second, by positioning themselves as superior with a consequent reduction of all others to a position of inferiority, available for exploitation. In their introduction to the *Book of Taliesin*, Gwyneth Lewis and Rowan Williams comment that these poems remind us in passionate, entertaining, challenging and, sometimes, humorous voices that we do not need to settle for such a drab, reductionist view of ourselves and the world in which we live.[18] Rather than see ourselves as superior and separate from the world around us, we can be open to the energy, vitality, enchantment and, at times, dangers of the world with its shifting, permeable borders and our human entanglement within it. We can reflect on what such openness can offer for living in today's troubled world.

It is hard to leave our willow seat at Rhyd Chwima, but it is time to do so and head 10 miles (16 km) downstream to Welshpool (Y Trallwng). Only 4 miles (6.4 km) from the Welsh–English border, the town is set on the western slopes above Hafren, at the point where Nant-y-Caws Brook joins her strong flow. This borderland town has a long history. Archaeologists have found evidence of Roman burial sites here as well as Roman coins and fragments of pottery. The town is believed to be built on the site of churches founded in the sixth century by St Cynfelyn and his brother Llywelyn. Records from 1253–4, refer to 'Capella de Trallu'g' ('The Chapel of the Pool Town'). This pool, thought to be either a large sheet of water located in what was once Powis Park or part of Hafren at nearby Pool Quay, was Hafren's highest navigable point in medieval times. In the twelfth century, Domen Gastell, a motte-and-bailey castle, was constructed close to Hafren, although from the thirteenth century, the mighty Powis Castle, which still stands today, became the main stronghold in the area. The town received a foundation charter in the 1240s, and records indicate that a market was held here as early as 1252. By the sixteenth century, the prefix 'Welsh' seems to have been added to the name of the town to distinguish it from other towns such as Poole in Dorset. The town remained strong as a trading hub, a position strengthened by the opening of the Montgomery Canal in 1796. The arrival of the Cambrian railway in 1862 reinforced the town's position as a regional centre. Today, Welshpool is at the intersection of three main roads and is well known as the entry point into mid Wales. The town's borderland position has also given rise to a curious entry in the *Guinness Book of Records* for twins born in different countries. In 1976, Carol, a local resident, did not know she was expecting twins. The first twin, Heidi, was born at Welshpool Hospital, but complications with the birth of Jo, the second twin, meant that Carol had to be transferred to the Royal Shrewsbury Hospital, 17 miles (27 km) away across the English border.

After our own less eventful journey we head into the town, keen to find something to eat. Many shops display their Welsh status proudly, such as the Welsh Gift Shop with its colourful display of arts and crafts produced by artisans from across Wales. We, however, are heading for the Little Welsh Bakery, particularly looking forward to the homemade bara brith it sells. Bara brith (*bara* – bread, *brith* – speckled) is a traditional Welsh 'baked good', existing in that space between bread and cake. Historically it was made from yeast dough enriched with sugar and dried fruit, a combination in many regional foods made by using up left-over dough as a little treat. These days, bara brith is most commonly made with flour and raising agent but is still made as a 'loaf', and slices are spread with butter. Its six basic ingredients are flour, dried fruit (soaked overnight in tea), eggs, sugar and a mix of spices. These may seem like common

Hafren

store-cupboard ingredients today, but each tells a story of newly introduced luxury items, often acquired through violent incursions into other lands. Spices were introduced during the Crusades in the medieval period. Sugar was also introduced at this time, although it remained a very rare luxury until the sugar trade, which developed alongside the slave trade from the seventeenth century onwards, began to make it widely available. John Burnett, a social historian whose work focuses on the working classes, estimated the annual consumption of sugar in 1801 was 13.87 kg (30.6 lb) per person.[19] The tea trade first came to Britain in the 1650s, but tea remained a luxury until the 1800s when it became more widely sold. Grapes have a more complex history; there are records of grapes being grown on the Welsh–English border as early as Roman times. Today there is still a vineyard at nearby Wroxeter, where the remains of a very large, strategically significant Roman settlement on Hafren's banks can be visited. That said, grapes and dried fruit would still have been rarely seen items until the trade in dried fruits became more widespread in the mid-nineteenth century.

Bara brith has entered into our own family folklore. On a family outing to the National Musem Cardiff, my sister Rachel and I decided that it was still too early for lunch, but we needed something to keep us going. We decided that bara brith – not quite as sweet or indulgent as cake – would be perfect. We asked Peter and James to buy some to go with our coffees. They didn't admit they didn't know what we meant, thinking they could ask for some at the counter. Their plan was foiled, however, when the girl at the till told them: 'Oh yes, help yourselves.' James and Peter cast around for clues as to what this mysterious bara brith could be, amidst much consternation as the coffee queue built up. Everyone in the family now knows and enjoys bara brith and can spot it in many a shop or café in Wales. I have made it at home, too, with a recipe given to me by a volunteer at our local hospital. This recipe adds orange marmalade to the other ingredients – their own family tradition, which I can highly recommend.

Bordering and crossing

The Little Welsh Bakery – its name painted in white on a black background above the narrow shopfront – is squeezed between a clothes shop and a former bank on Welshpool's busy Broad Street. The window is crammed with delicious-looking loaves of bread and savoury and sweet bakes – a seventh heaven for Peter. Entering the shop, we can see different sizes of bara brith on the shelves behind the counter. The welcoming two young women working there advise us on our purchases – a traditional savoury pie each with leek and cheese filling and several bara brith loaves, dark and filled with plump fruit. We also buy coffees to have with our picnic and, well laden with treats, set out to the Montgomery Canal, which passes through the town centre.

The canal is unusual as it was originally built in the 1790s to support agriculture rather than industry, transporting limestone quarried at Llanymynech to canal-side kilns along its route used to transform it into quicklime to spread on fields. This reduced soil acidity and improved crop yields at a time when the increase in corn (wheat) prices, caused by the ever-increasing demand for flour from urban populations, made growing crops more attractive, even on land less suited to this use. However, the depressed state of agriculture in the second half of the nineteenth century, together with increased use of rail transport and the availability of alternative fertilisers, led to a fall in the canal's use. By the 1890s, the canal was barely covering the cost of maintenance. It struggled on but fell into disuse in the mid-twentieth century. However, an 8 mile (12.9 km) section, with Frankton Locks at its northern end where it joins the popular Llangollen Canal, has now been restored after a lengthy project.

Today, the section of the canal in Wales is designated as a Special Area of Conservation (SAC) and Site of Special Scientific Interest (SSSI) due to rare plants such as the floating water plantain, *Luronium natans*. The canal brings nature into the heart of Welshpool, attracting otters, kingfishers, dragonflies, damselflies and water voles. We make our way down onto the towpath at Canal Wharf and head eastwards. The bustle of the town soon fades,

Hafren

and we settle down to eat our picnic – absolutely delicious. As we listen to the sounds around us – the distant gentle hum of the town blending with the birdsong – we watch as ducks and two swans glide by. We are lulled by the peacefulness around us, refreshed and nourished by both the food and this place: the glassy stillness of the canal water, the greens and browns of the trees, the occasional bright pink splash of summer flowers and the frothy white hedge parsley growing along the banks.

Restored, we leave Welshpool and travel the 4 miles (6.4 km) to cross the now-invisible border between Wales and England. 'ARAF' changes to 'SLOW' on the road markings as we approach Shrewsbury – a Marches town existing in that shifting space between ancient Welsh and English kingdoms. Its medieval centre is almost entirely encircled by Hafren, and the small aperture in Hafren's course is protected by a castle on a high bluff. Made of imposing red stone, the castle was first built as a timber fortification in 1070 by Roger de Montgomery. Vehicles can now cross into Shrewsbury's centre using the Welsh Bridge on the western side and the English Bridge on the eastern flank. It was there that Roger de Montgomery founded an abbey in 1083 on the site of an existing Saxon church. Together, the abbey and castle were key to Shrewsbury's growth into a significant and powerful market town by the medieval period, with, for example, its Drapers Company controlling the trading of the flannel produced in mid Wales. With a population today of approximately 72,000, Shrewsbury continues to be a popular market town. In the evening light I stand on the Welsh Bridge, looking at Hafren flowing swiftly through the stone arches below my feet – lively, powerful, holding both danger and potential. I reflect on Welsh stories and traditions of liminal, in-between artefacts and spaces, as well as shifting borders, including the shifting of Hafren's own role as a border over the centuries. These remind us that the hard borders we build between ourselves and the wider natural world can be crossed, can change, can be transformed.

Chapter 5:
Meandering
Shrewsbury, Shropshire

Meandering is a very 'rivery' action. In Greek mythology, Maiandros is the patron deity of the Menderes River, well known for its sinuous curves as it flows through what is now modern-day Turkey. Hafren meanders, too, wandering without urgency, frequently changing direction, circling, travelling indirectly. As a mountain stream in Pumlumon, Hafren flows rapidly, the forceful water carving into the rock causing vertical erosion. On the lower land, which begins after Llanidloes, she flows more slowly, and her meanders form due to horizontal erosion. This begins when an initial disturbance on her bank, such as animals burrowing, weakens that area and it eventually collapses. This opens up a space for water to flow in, which erodes the bank further and the hole deepens. More water rushes in, speeding up the flow of water towards this bank, and the subsequent hollowing-out forms into a curve. As the current strengthens towards this bank, the flow weakens near the other bank. Since this slower water does not have the energy to transport as many sand-sized particles as the faster water, these particles fall to the bottom of this slower side and eventually new land forms. In addition, when the faster water hits the bank on the fast-flowing side, it ricochets towards and hits the other bank with force; this then becomes the next place where soil is eroded. The process is repeated and, eventually, alternating meanders form. The wider a river, the longer it takes for this ricocheting current to reach the other bank, so the distance to the downstream curve is longer. Studies show that this distance is remarkably regular, with the distance between meanders being about six times the width of the river.[20] Very tight meanders, such as the one around Shrewsbury, form when silt increasingly builds up on the 'slower side'. Left to their own devices, rivers with such tight meanders would eventually break through the land at this point, especially during flooding. The course of a river then straightens out and an oxbow lake is left behind.

Such natural forces are never far from the minds of those living and working in Shrewsbury. When we return to the town the following spring to spend a night held tight by Hafren's watery loop, we see sandbags piled high in readiness in an alleyway off the Mardol. Hafren had recently flooded, her sediment-rich swirling water breaking over her banks as well as rising up through the cellars and drains. The water has receded now, the town returning to a drier state. After a peaceful night, we awake to a cold, overcast but, thankfully, dry day. Keen to 'walk the loop', we set off through the town. The main street rises upwards towards the castle mount, protecting the narrow stretch of land that joins Shrewsbury to the surrounding Shropshire countryside. This town holds a special place in my heart. It is here that I bought my wedding dress on a shopping trip with my mother almost forty years ago and the man I

Meandering

married walks beside me now. We stop at a bakery, attracted by the Shrewsbury buns in the window and learn that these are made from dough enriched with butter and eggs. Peter loves a sweet treat, so we buy two to enjoy later beside Hafren.

As we pass the gardens in front of the town's library and museum, we see tributes to two of Shrewsbury's famous citizens. In a lawned area surrounded by flowers, we admire the bronze bust of local poet and novelist Mary Webb, her head and shoulders emerging from a stack of books. Set on a plinth of creamy coloured local Grinshill stone, the bust was placed here in 2016 in front of the library where she borrowed books and attended talks. Taking centre stage, however, is a large bronze statue of Charles Darwin, seated in an armchair with a pile of books at his feet, set on a high plinth of polished black granite. Darwin is world-renowned for his insights and theories on natural history, which completely changed understandings of the origins and diversity of the planet's species. His ideas have become commonplace in everyday life and integrated into scientific, literary and many people's religious beliefs. However, they were originally considered by many as radical, even blasphemous, as his theories challenged literal interpretations of biblical creation stories as well as other beliefs.

Darwin was born in 1809 into a prosperous Shrewsbury family. Their home, Mount House, stood in extensive grounds sloping down to Hafren on the western flanks of the town beyond the Welsh Bridge. The fields, hedges, trees, wildlife and water surrounding his home nurtured his early interest in natural history and geology and it was here that he first collected plants and insects. His enquiring mind was encouraged in a family considered to be 'free-thinking' – holding the view that understanding should be formed on the basis of empirical observation, logic and reason rather than dogma or the authority of others. Following in his father's and grandfather's footsteps, Darwin went to study medicine in Edinburgh in 1825 but found the blood and brutality of medicine at that time very distressing. During this time, he learnt the skills of taxidermy, which were to prove useful later in his career. In 1828, he changed direction to study theology at the University of Cambridge, where he also continued his interest in natural history. There, he was recommended by one of his professors as naturalist and companion to Robert FitzRoy, captain of HMS *Beagle*, on a journey to survey the coast of South America that would last five years.

During his travels, Darwin experienced a huge diversity of plants and creatures, including giant turtles, vividly plumaged birds and sharks of different shapes and sizes. He took copious, meticulous notes and collected specimens, using these as the basis for his theory of evolution, which he developed over the next twenty years. Other pioneers of biology and geology were also developing similar ideas; the Welshman Alfred Russel Wallace had independently developed a theory of evolution. The two men jointly presented their ideas on evolution to the Linnean Society in 1858 and Darwin published *On the Origin of Species* in 1859. He argued that species survive through a lengthy process of 'natural selection'. Those that adapt to their environment survive and thrive and those that do not die off (over hundreds of thousands of years). For example, in the Galapagos Islands, off the coast of Ecuador, Darwin saw finches on some of the islands that ate nuts and seeds. These had adapted to have stout beaks. On other islands, finches were insect-eaters and had evolved to have thin, sharp beaks adapted for this. In total, Darwin identified thirteen different types of finch, all descended from a single ancestor. Very controversially, Darwin also proposed that humans belong to the same 'great apes' family as chimpanzees, gorillas and orangutans, all descended from a shared ancestor.

In contrast to Darwin, Mary Webb, poet and lyrical novelist, has a quieter story to tell. Born in 1881 in the small village of Leighton, 10 miles (16 km) south-east of Shrewsbury, she was of Celtic descent on both sides of her family. Her mother was Scottish and her father, a gentlemen tutor and nature lover, was proud of his Welsh descent. The eldest of six children, Mary's early creative output was stories and plays to entertain her brothers and sisters. She also loved to roam in the Shropshire countryside and developed a special bond with the wider natural world that lasted throughout her life, pining for the Shropshire borderlands whenever she was away from them. She had a particular lyrical talent for describing the intricate details she experienced. For example, in her 1917 essay 'Vis Medicatrix Naturae' she describes the 'soft grey cloud, where one lone minstrel thrush is chanting to the dying light' and how the 'delicate traceries of silver birches are tenderly dark on the illuminated sky'. Her special connection with nature was particularly important to her when, in her early twenties, she developed Graves' disease, a condition that damaged her thyroid gland. This affected her throughout her life and eventually caused her death in 1927 at the age of just forty-six. The disease also altered her appearance, which made her feel self-conscious, and she found that time in nature was a particular solace. It was during the first flare of this illness that she wrote 'Vis Medicatrix Naturae', describing her bond with nature as 'the sudden sense – keen and startling – of oneness with beauty seen and unseen'. It is in such moments that we 'know that we are not merely built up physically out of flower, feather and light, but are one with them in every fibre of our being'. For Mary:

> Life, the unknown quantity, the guarded secret, circles from an infinite ocean through all created things, and turns again to the ocean. This miracle that we eternally question and desire and adore dwells in the comet, in the heart of a bird, and the flying dust of pollen. It glows upon us from the blazing sun and from a little bush of broom, unveiled and yet mysterious, guarded only by its own light – more impenetrable than darkness.
>
> The power of this life, if men will open their hearts to it, will heal them, will create them anew, physically and spiritually.[21]

Hafren

Her lyrical novels, the most acclaimed of which is *Precious Bane*, published in 1924, evoke the landscape and people of her beloved Shropshire. She received posthumous acclaim after the then prime minister, Stanley Baldwin, praised the work of this 'neglected genius' in his speech at the Royal Literary Fund Dinner in 1928. Her work is again receiving attention as awareness grows in Westernised capitalist societies of the need to find ways to experience and live in relationship with the world around us – something many other cultures have long understood.

Particularly affected by Mary Webb, the sensitivity of her work and her connection with the wider natural world, we linger before her bust. Then, turning slightly right, we walk towards Shrewsbury Castle, its crenellated red walls built of sandstone. Now a military museum, it is closed today, but from its ramparts we have a view over the tightly packed lower town. Next to the castle is Shrewsbury's grand railway station, a castle of its own, built to impress and proudly announce the importance of the railway in the Victorian era. Continuing along a road that passes under the railway station, we move outside the loop enclosing Shrewsbury's medieval centre and become slightly lost in an industrial area. Peter jogs ahead and, spotting Water Lane and Severn Lane, calls out: 'Over here.' Sure enough, at the end of a sloping, narrow street flanked by two large houses with high, red-brick garden walls, we can see a framed view of Hafren's grey-green water set against a backdrop of tall, bare trees and gently drooping willows on the opposite bank. As we reach the pedestrian walkway that runs alongside Hafren's inner bank, I feel space open up before me and take a deep breath in, then out. My shoulders relax as I adjust my eyes to the longer views now afforded to us in both directions. To our left, we can see Hafren flowing downstream away from the town. A sign hangs from a pedestrian bridge warning of the weir situated just beyond the bend. The bare trees fringing the bank opposite create inky reflections in the water, and the occasional glimpses of sunshine, breaking through the overcast sky, cast golden pools of light.

Meandering

We turn to our right and begin our walk around Hafren's loop. On this inner bank, houses are set high above us, their steep gardens ending in high, red-brick retaining walls. Rather than succumb to the frequent flooding, the town has adapted. The path beneath our feet is covered in creamy-coloured silt residue, which also extends about 1 m (3 ft) up the retaining walls, indicating the height of the recent sediment-rich flood-waters. The brambles and grasses beside the water have a frost-tinted, wintery hue – a bizarre sight on this spring day – but again, this is silt. Men wearing council waterproof high-vis jackets are clearing away signs saying, 'Footpath closed due to flooding'. After walking under the railway bridge, we find a sunny bench where we sit for a few minutes to watch the rapid, swirling water and the passers-by: joggers; an older couple out strolling; two mothers with young children on scooters; a dad with a pushchair, his smiling toddler waving her arms. We taste the Shrewsbury buns we bought earlier and find them light, sweet and buttery – delicious. Refreshed, we set off again and soon pass under the English Bridge. A bridge is known to have stood here since the eleventh century. This latest version, built in 1926 to cope with the increased traffic, includes reused stones from the bridge built in 1774. Branches, brought here by the recent powerful floodwaters, are caught up around the piers supporting the arches. Animal tracks criss-cross mud the colour of milky coffee. I wonder if these are from dogs, badgers or other creatures?

We pass under the metal fretwork of the Greyfriars footbridge, busy with pedestrians making their way into Shrewsbury's medieval centre. On the banks, several anglers are making use of the small platforms set into the bank by the town council. We do not want to disturb their contemplation but seeing one about to pour a coffee, we approach and, after checking if it is ok to speak to him, we ask what he enjoys about coming here. He replies that just as important as the fishing, is the opportunity to be outside, beside the water, to enjoy the changing seasons and the variety of plants, birds and animals these bring. When I ask what fish swim in the water here, I am very surprised by his long list. There are roach, perch, gudgeon, minnows, chub, carp, dace and salmon, although, he comments: 'Salmon do not tend to linger long in the loop.' There is so much life we cannot see beneath the surface of the water. We pass tennis courts and a bowling green, and on the opposite bank, we can see the rowing and boating clubs as well as houses built high above the flood line with terraced gardens leading down to the water. These often include a chair or two, to enjoy the passing scene during more clement weather. We soon enter a parkland area called the Quarry Park. This 29-acre (11.74-hectare) green space was established in 1719 in the town centre. It has been retained to this day for pleasure, and to accommodate the floodwater that spreads outwards from Hafren here on the loop. We walk across grass, still spongy from the recent flooding, towards the Dingle, an ornamental garden opened in 1879, hidden behind beech hedges and railings in the middle of the park. The garden had started life as a 'wet quarry', providing poor-quality red sandstone for construction in the town, but now it is filled with flowers and shrubs just beginning to reveal their spring colour.

In a corner, under an ivy-clad, deep arch, we find a pale classical statue of a semi-naked Sabrina (the Latin name for Hafren) nestling in a pool rippled by a jet of water that rises beside her. The attractive statue was created by Birmingham artist Peter Hollins and presented to the people of Shrewsbury by the Earl of Bradford in 1846. Lounging sideways, with one hand in her flowing hair, she looks down at the water in contemplation. The liveliness of the water, combining with Hafren's contemplative expression and beauty, draws us in. On the plinth we read these words from *Comus*, a masque (an enacted poem) by John Milton:

> Sabrina fair,
> Listen where thou art sitting
> Under the glassie, cool, translucent wave,
> In twisted braids of lilies knitting

Meandering

The loose train of thy amber-dropping hair;
Listen for dear honour's sake,
Goddess of the silver lake,
Listen and save.

First performed in Ludlow Castle, Shropshire in 1634, *Comus* is a fable of the battle between virtue and vice, good and evil. It tells of the quest undertaken by two brothers to rescue their chaste sister (called The Lady in the poem) from the evil clutches of Comus, the god of revelry, who has transformed himself into a villager. Comus has trapped The Lady in an enchanted chair and is trying to get her to drink from the cup that represents sexual pleasures and over-indulgence. But The Lady refuses, citing the virtues of chastity and rational self-control. The brothers are helped in their quest by The Attendant Spirit, who has transformed himself into a shepherd. He tells the brothers how to defeat Comus and also sings the song engraved on the statue to summon Sabrina (Hafren) to free The Lady. Sabrina is able to do so as The Lady has not drunk from the cup. Thus, goodness and virtue triumph over vice and evil. The siblings are reunited and return to their family for celebrations. The main themes of this masque are straightforward, with stereotypical roles for men and women. It does, however, also contain more nuanced arguments around gender, and the lively conversations between The Lady and Comus challenge stereotypes around the rational capabilities of women.

Taking a last look at this statue of Hafren, sad to leave her, we head onwards through the Quarry Park to the waterside outdoor café set beside a children's play park. Today, the café is firmly on the ground, but its base conceals hydraulic 'jacks', which raise the café above the water level in periods of flooding. I had seen recent pictures of the café appearing like a wooden boat floating on a vast expanse of water. We are pleased to see it open today, serving some much-needed coffee. The play park seems none the worse for its recent soaking, and children are happily clambering and sliding.

Hafren

Tired now, but a little restored by our coffee and rest, we set off once again through the park to complete the final stretch of the loop, passing beside the attractive metalwork of Port Hill pedestrian suspension bridge and finally the stone-built Welsh Bridge.

In the Mardol Quay gardens beside the bridge we stop to look at the 12-m- (39-ft-) high concrete, curved sculpture reminiscent of a dinosaur's spine. Called *Quantum Leap*, it was unveiled in 2009 to commemorate the bicentenary of the birth of Charles Darwin. Beside the sculpture is a bronze plaque that depicts a timeline of the geological periods of the earth's four-and-a-half- billion-year history, pictorial representations of living organisms from these periods based on fossil records, and information about extinctions that have occurred. What really strikes me is that the oldest member of the genus *Homo* depicted here, called *H. habilis*, lived a mere 2.3–1.4 million years ago, and his widespread descendant, *H. erectus*, lived from 1.9 million years ago to 100,000 years ago or maybe even later.[22] This visual representation, albeit using a simplified timeline, greatly emphasises just how recent humans are in the history of the planet.

Whilst such understandings of time and the age of the earth and its inhabitants are widely accepted today, this was not always the case. In the seventeenth century, the Anglo-Irish archbishop James Ussher calculated the age of the earth by adding up the generations mentioned in the Bible, coming up with a figure of 4,004 years to get back to the creation story. However, through observing the natural world and geological records around them, many scientists and philosophers across Europe in the seventeenth century began to challenge ideas based on such biblical sources, and their alternative views were seen by many Christians as blasphemous. The ideas of these dissidents were drawn together and persuasively communicated by Charles Lyell in his three-volume *Principles of Geology* published in 1830–3.[23] It is well documented that Darwin took of copy of Lyell's book on his journey on HMS *Beagle*. These ideas gave Darwin the 'gift of time' needed to support his theory of evolution. If plants and animals survived through adaptation and different branches developed from a single genus, this required an understanding of time that allowed hundreds of thousands if not millions of years for these biological processes to occur.

As we walk the final stretch of the loop back towards the castle, we pass in front of a large white pub prone to frequent flooding. This town has existed here for almost 1,000 years and the people have found ways to cope. The owners of this pub have now carpeted the bar and restaurant with water-resistant Astroturf, set electrical plugs high in the walls and raised the cellar contents on plinths, since floodwater often rises up through cellars in the town as well as overflowing Hafren's banks. Other residents have built houses high above the water level, protected buildings with flood walls and left open space, such as the Quarry Park, for floodwaters. On this walk I expected to see evidence of a town fearful of flooding, but instead I have had a sense of their enjoyment of the walks along the riverbanks, the green spaces for relaxation and the various land and water sport facilities. Flooding *is* problematic and costly for Shrewsbury, with an increasing number of significant floods occurring. A local shopkeeper told me that five of the highest floods ever experienced in Shrewsbury have been in the last fifty years. Yet the town persists and takes pleasure from its proximity to Hafren.

After another night tucked within Hafren's loop, it is time to leave Shrewsbury, and we set off eastwards over the English Bridge. We pass the place, just outside Hafren's loop, where Roger de Montgomery founded Shrewsbury Abbey in the eleventh century, although the site is now occupied by a much more modern building. We then turn left towards the house where the poet Wilfred Owen once lived. Owen was born at his mother's family home near Oswestry, Shropshire in 1893. After a period in Birkenhead, the Owen family moved to Shrewsbury in 1907 so that his father could take up his post as assistant superintendent of the railways, living at various addresses before renting 69 Monkmoor Road. We stop in front of the semi-detached, red-brick, Edwardian three-storey house, taking in the simple blue commemorative plaque stating: 'Wilfred Owen, poet, lived here, 1910 to 1918'. The house is still called Mahim, the name given to it by Owen's father, Tom, as a souvenir of his time in India. We look up at the dormer window of the attic that Wilfred shared with his brother, Harold, which had, at that time, open views across to the old racecourse. The family used to walk across nearby meadows on Sundays to the ferry which linked Monkmoor to Uffington, where they attended church. An Owen family story tells of how the pollen of the buttercups that grew there adorned their boots with gold. This finds an echo, some say, in Owen's poem 'Spring Offensive' set on the Western Front. In this poem he describes a scene in which the soldiers:

> Hour after hour [they] ponder the warm field –
> And the far valley behind, where the buttercups
> Had blessed with gold their slow boots coming up

Owen was a scholarly boy, talented at writing even at a young age. He desperately wanted to continue his education and attend university but his parents were not in a financial position to support this. Instead, he was enrolled in the Shrewsbury Technical School to train as a pupil-teacher. He continued his more academic studies alongside this and his teachers encouraged his interest in English literature, especially poetry. Owen set up a table and chair, which he called 'The Study', in his attic bedroom. He then worked as a pupil-teacher at Wyle Cop School in the town. He was reluctant, however, to take the next step of training to be a teacher since he would then have been required to commit to teaching in state-funded schools for at least seven years. Unable to obtain a scholarship to attend university, he became an assistant in a parish, with time allocated for preparing to repeat the scholarship examinations. He next taught in France before volunteering in 1915 to fight in the First World War. He served in the trenches in France before spending a year in Edinburgh recovering from shell shock. This is where

Meandering

he met Siegfried Sassoon, who had a profound effect on him and drove him on to work ever harder on his poetry. What strikes me, reading about Owen, is the sheer 'hard graft' and dedication he put into his poetry. In his final months, before he returned to France in 2018 and ultimately to his death, he hired a room in a cottage so that in his spare time he could work in solitude away from the army camp. My impression when first learning about Owen when I was at school was that his poetry was a spontaneous, of-the-moment response to the horrors of war. His toil, his internal struggles over his sexuality, his fears over the inadequacy of his poetry, his thwarted striving to finance a university education, never received attention – a misrepresentation and a missed opportunity.

Owen's struggles were very much on my mind as we walked the short distance from his home on Monkmoor Road to Hafren's banks to see the weir. Constructed in 1909, this was built to keep water at a height suitable for pleasure boating and promenading as Hafren passes around the town. Above the weir, the water is still. Willows lean over this mirror-smooth, steely expanse, their reflections creating dark, silent sentinels. But then, the water roars over the weir, falling into a turbulent mass below. The roar matches my own troubled thoughts. I wondered if Owen made this walk himself, watching Hafren – released from the loop – flowing swiftly, noisily away from the town.

Chapter 6:
Industrialising
from Shrewsbury, Shropshire, to Ironbridge, Shropshire

Like Hafren, Peter and I leave Shrewsbury and head downstream towards Buildwas where the ruined Cistercian abbey raises its arches and crumbling walls against a backdrop of dark trees. Just after the village of Leighton, we go around a bend in the road and the view suddenly opens up. From our raised position, we can see verdant fields, dotted with trees and grazing sheep and lambs, gently sloping down to Hafren, her curves winding through the peaceful valley floor. We stop at the Severn viewpoint to enjoy this particularly meandering section created by the changing energies and interactions between land and water. Hafren is flowing calmly between her banks today, but in the winter months she fills the valley bottom here with rippling water, trees and hedgerows seeming to float in the silvery lake that has formed around them. As the waters recede after each flood, how long will it be until Hafren finds a straighter course, cutting off these curves, leaving oxbow lakes behind as markers of her past route?

I am keen now to meander alongside Hafren, but there is no footpath here, so we return a short distance back upstream to Cressage. This village has an interesting history, taking its name from *ac*, the Anglo-Saxon word for oak: it is reputed that St Augustine of Canterbury preached here in 584 CE under an oak tree. In the Domesday Book, a survey of Britain undertaken by order of William I in 1086, the village is recorded as Christache, meaning 'Christ's oak'. Arriving in the village, we walk to the war memorial that stands at the place where legend says the Cressage Oak once stood. This well-kept monument to the village's own 'fallen oaks' is a simple stone Celtic cross on a tall plinth. Four small laurel bushes in planters break up the drabness of the grey slabs surrounding it. After some moments of reflection, we walk northwards to the balustraded bridge and, crossing over a stile, we drop down onto the path running beside Hafren. A mix of grey, white and weeping willows form soft mounds amongst the tall grasses that rustle in the breeze. All is reflected in the still water. Two swans glide by. Peace descends on my spirit as we sit down on the bank to rest. Walking on, we find a tall oak, the now-bare branches on one side spreading out over the grassy bank and water. Here we can imagine St Augustine holding forth, the oak's canopy providing a perfect shelter from summer sunshine or rain.

At first, the path follows a meandering course beside the water, but then it rises and enters a thicket of trees, predominantly hawthorn with small signs of the delicate leaf buds almost ready to emerge. Glancing down, I can see Hafren 6 m (20 ft) below us, meandering around a large bend. The path narrows and the tightly packed, intertwined hawthorns increasingly catch hold of us as we squeeze between them. As I look closely at the branches, their emerging, tender growth contrasting with the spiky thorns, I think of Mary Webb's poem *Green Rain*. In this, she describes going into the 'scented woods' where:

> There are the twisted hawthorn trees
> Thick-set with buds, as clear and pale
> As golden water or green hail –
> As if a storm of rain had stood
> Enchanted in the thorny wood,
> And, hearing fairy voices call,
> Hung poised, forgetting how to fall.

Reluctantly, we decide to turn back, but these bushes cling tight to us, catch our clothes, our hair. We are entangled, enmeshed. It takes a while to unpick ourselves and, despite the pain I feel when the thorns scratch my skin and tug on my hair, I feel happy. We can both laugh at our predicament, at this unexpected encounter. Eventually we pull ourselves free and, somewhat sadly, manage to turn and walk back without too much further ado, although the hawthorns do have a few final attempts to hold us firm in the woods.

Meandering with Hafren today has encouraged us to slow down, take a few different turns here and there, be taken by surprise, be caught up. This is something we often lose sight of in modern society with its emphasis on travelling from A to B as quickly as possible. However, as anyone who has tried to meditate knows, it seems that the human mind constantly seeks to meander, to be more like Hafren and the desire she expressed in the giant's myth to seek out twists and turns and spend time with all she meets along the way. Researchers at the University of California, Berkeley[24] found that babies' and young children's brains seem to wander constantly, and they wondered what purpose this served. By monitoring brain activity using electrocardiogram (ECG) measurements, they were able to identify different patterns that occur when the

Industrialising

brain is task-focused, freely wandering or constrained. Based on these observations, they argued that focusing only on the task in hand can lead to missing key information. In contrast, when free-association thought processes are encouraged, they randomly generate memories and imaginative experiences that can interact and draw out new insights and ideas – something very much needed in the world today. Physically meandering is important, too. It exercises our bodies, reduces stress hormones and can contribute to improved mental health. Meandering whilst using all our senses to pay attention can open up space and time for new experiences and unexpected meetings. This can contribute to the making of lasting memories, which can, in turn, be springboards for action and care. From my own life, I can recall many such special moments: an encounter with a roe deer at Minsmere in Suffolk, raising her head and gazing towards me, immobile, haunting; an encounter with damselflies hovering over the Thames as I meandered through a flower-filled meadow in Oxfordshire; an encounter with an otter smoothly, determinedly swimming up-river from a sea loch in Kintyre, Scotland; this encounter today with hawthorns in Shropshire.

Peter and I continue downstream to visit Ironbridge, where the world's first iron bridge was built by the Darby family, to join Hafren's steep-sided wooded banks. I am excited to have my first glimpse of the famous bridge standing tall and proud, its high arch designed to let cargo boats pass through in the days when Hafren was part of a busy transport network. The year it was erected – 1779 – is picked out in white on its main arch, its delicate, ornate metalwork belying its strength. I was expecting the bridge to be black as in my childhood recollections, but it is now a deep reddish-brown, the original colour, we discover. The town's bold claim is that it is the 'birthplace of industry'. Today, my understanding and concerns over the problems created by the European Industrial Revolution, not least its contributions to climate change and colonisation, have grown. However, when I was a child, we were

Hafren

proud of this idea when we took our visitors to see Hafren and the famous bridge. We stop to look more closely at the bridge and take a photo together to match the one in my family album, where I am standing with my mother, my sister and our friends. Given Hafren's reputation for flooding, it seems particularly appropriate that in that photo I am holding my much-loved 'Rainy Day Doll' dressed in her short yellow sou'wester coat and hat and white wellingtons, their colour so reminiscent of the 1960s. I am happy to say that I still have this doll. The plastic of her coat and hat is a little crisp now but her neat fringe and cheeky smile provide a happy link to this place on Hafren's banks, still vivid from my childhood. On our visit today, we are surprised to hear so many languages around us: French, German, Chinese, and also Portuguese, spoken by a large group from Brazil manoeuvring themselves into a position for a photograph in front of the famous bridge. This place, it seems, still has the power to draw people in.

We cross the bridge to look inside the museum housed in the booth where a toll was once charged. For me, Hafren and Ironbridge are intimately and intrinsically linked. I have crossed this bridge many times before, holding my mother's hand, watching the fast-flowing water far below, and eaten many an ice cream with my sister on Hafren's banks. These memories meld together in my mind. It seems that Ironbridge and Hafren hold a special place in other people's memories, too. In the small museum I notice that Peter has paused in front of the reproduction of a painting. I go over to him and look. It is an oil of the Bedlam Furnace, built on Hafren's banks a mile downstream from here so that the large pieces cast at the foundry could be transported more easily by water. Called *Coalbrookdale by Night,* Philippe Jacques de Loutherbourg painted it during a tour he made between 1799 and 1800 to record the industrialisation of England at that time. The original is now displayed at the Science Museum in London. This depiction has become a symbol, for many, of the birth of the Industrial Revolution, and all the power that it unleashed, here at Ironbridge and in Coalbrookdale, a side valley

Industrialising

off the main gorge. The night sky and surrounding buildings are lit up with billowing gold and orange light emitting from the hellish furnace, as men and horses work tirelessly through the night in the foreground to haul raw materials to the furnace and finished pieces to Hafren. 'This picture was in my primary-school history textbook,' Peter tells me. 'It takes me right back. I remember thinking Britain was all about sheep and monasteries but then mass iron production started, and it became all about industry.' He explains how he is just realising how much of his feelings about Britain's place in the world are still shaped by this painting. Peter isn't the only person I know whose history books featured Hafren and Ironbridge. A friend recently told me how an image of the iron bridge, pictured on the front of her school history book, sparked her imagination and love of history. It was a very striking and moving moment for her when she finally saw the bridge for the first time many years later.

These innovations, celebrated in those schoolbooks, were thanks to the energy, originality and observations of Abraham Darby the Elder. To learn about these innovations, we make our way to Coalbrookdale Museum of Iron, high above the town. Born in 1677, Abraham Darby spent his early years learning the brass foundry business where he saw molten brass poured into moulds (cast). He experimented with green (damp) sand moulds, which enabled him to cast brass pots and cauldrons larger and thinner than those cast in the loam moulds traditionally used at that time. During this early period of his life, he also observed the innovation of using coke (made from coal) rather than charcoal (made from wood) to fire the malting ovens used in the brewing industry. In 1708, he leased a furnace in Coalbrookdale, and, combining his various observations, set about developing a blast furnace to mass-produce cast iron. Darby could find all the things he needed here. Iron has been extracted from iron ore for millennia, including in Coalbrookdale, but the reliance on charcoal made from wood limited how much iron could be produced. By using coke produced by charring coal in sealed chambers, the Coalbrookdale works were able to produce high-quality iron at scale. The readily available local coal, with its low sulphur content, made good coke.[25] Sand needed for moulds, and limestone, to remove impurities, were also available, and the streams running rapidly down the steep-sided gorge provided water power. Hafren provided a means of transport.

Three viewing levels have been created at the museum around Darby's original, but now silent, furnace to help visitors understand how iron was extracted from iron ore using 'hot chemistry' at temperatures of up to approximately 1,500°C (2,731°F). We climb to the top platform and look down through the 'mouth' into the brick-lined furnace. The process starts here when crushed coke, iron ore (containing iron in the form of iron oxide) and limestone are added in a continuous flow. The furnace is essentially a huge melting pot where solids sink and gases rise. When blasts of oxygen are introduced to set the charcoal to burn, a chemical reaction produces carbon monoxide that 'grabs' the oxygen present in the iron oxide, separating it from the iron part of the compound, leaving molten iron. The limestone also goes through a number of chemical reactions that lead to the separation of impurities in the iron.

At the second platform, we peer inside the blackened furnace where these processes once took place. I fancy I can still smell the distinctive tang of metal I had previously encountered in smithies and metal workshops. Finally, descending to the ground-level platform, we can see we can see the small opening at the base of the furnace, with Darby's name inscribed above, where the molten iron once flowed out: a dangerous place of intense heat and brightness. The impurities (slag), lighter than the iron, float on top and are skimmed off. The molten metal is then channelled into sand moulds with a main runner on one edge and side moulds coming off this, resembling a sow (the runner) feeding her piglets (the side moulds) – hence the term 'pig iron'. Alternatively, the molten metal could be channelled into large moulds developed by Darby and his workers so that a variety of products could be cast. Darby's practical understanding of these processes is even more remarkable

Industrialising

when one considers they were developed before the individual chemical elements used in this explanation were identified.

At first, the dependence on water power meant that in summer the local streams ran too low to power the bellows used to introduce oxygen into the blast furnaces. Instead, this period was used for maintenance. However, the development of steam-powered engines to drive the bellows and the seemingly endless potential of coal made the mass production of iron possible and cheap. Further furnaces were built by the Darby family and other local businesses, including the Bedlam furnace, located on Hafren's banks for easier access to river transport. The iron mass-produced in Coalbrookdale was used in the eighteenth century throughout Britain for the production of a whole host of crucial innovations including iron cylinders for steam engines, iron wheels for railways wagons, iron rails, iron river barges and, of course, the now-famous iron bridge.

River transport continued to be important during this period. A special boat, the shallow, flat-bottomed Severn trow, had developed over many centuries to suit the particular conditions on Hafren and her sister river, Gwy. The word 'trow' is thought to come from the Saxon word *trog* meaning 'trough-shaped container'. There were two main types. The smaller up-river trows, normally with a single central sail across the boat, like a Viking longboat, operated on routes as far up-river as Shrewsbury and Welshpool. The larger, downstream trow sailed between Worcester, Gloucester and the Bristol Channel to ports such as Newport and Cardiff. A log of wood could also be strapped to these larger boats to provide a temporary keel to make the trows seaworthy. They had fore and aft sails that pointed along the boat and masts that could be lowered to pass under bridges. They did not have a deck except over small cabin areas at each end; the cargo was protected by canvas bulwarks. This served to keep the weight of the trow down, which was important in the summer months when the water was low and teams of men had to pull the boats over the shallows. The *Spry*, built in Chepstow in 1894, is the last-known surviving Severn trow. In 1982 she was rescued from the Diglis Basin in Worcester and restored to a river-worthy condition. After a number of expeditions and events, she was taken to the Blists Hill Victorian Town here at Ironbridge where she is displayed in a specially built hangar.

Coalbrookdale began to decline in the nineteenth century. It could no longer compete with foundries and manufacturing in Staffordshire, where advances in mine engineering made the deeper coal seams there easier to mine. The transport offered by Hafren could not compete with canals and the railways. Darby's legacy, however, lives on today – for good and ill. Steel, which is made when iron is further refined and additional minerals are added, is a key material in the present-day construction and engineering industries. Drawing on stores of buried energy (fossil fuels such as coal, oil and gas) enabled industry to expand rapidly. This is very widely believed to have had devastating consequences for the delicate ecosystems on which life on earth depends. Levels of carbon dioxide in the atmosphere have increased by over 140 per cent compared to pre-industrial levels. Methane levels have increased by over 260 per cent and nitrous oxide levels by over 120 per cent. There is a general scientific consensus that these gases, known collectively as greenhouse gases, contribute significantly to global warming since their presence traps an increasing amount of heat from the sun in the earth's atmosphere. Also very significant is that these gases are increasing in the atmosphere at an ever-increasing rate.[26]

As the sea warms, it expands. When combined with melting polar ice, this increased volume of seawater threatens the very existence of low-lying coastal areas and land bordering estuaries such as land along Hafren's estuary as she enters the Môr Hafren (Bristol Channel), as well as entire island communities and ecosystems. Warmer seas also increase evaporation, leading to destructive storms and flooding. Higher temperatures are causing more frequent heatwaves, leading to water and food shortages, health issues, increasing desertification, increased poverty and the displacement of peoples. Species loss is exacerbated by drought, forest fires, human

Industrialising

encroachment and the movement of some species into areas where other species have no resistance to them (and have insufficient time to adapt). The world is losing species at a rate 1,000 times greater than at any other time in history. One million species are at risk of extinction within the next few decades. Many of the effects of global warming are happening in areas which have not created it, leading to calls for climate justice. Industrialised countries are being asked to provide support (including financial) to mitigate current and future effects of increasing world temperatures, as well as to make plans to reduce current and future carbon emissions.[27]

Faced with this damning information, it is easy to despair, but other responses are both possible and necessary: responses that focus on restoring rivers and the ecosystems in which they are embedded. In many traditional understandings and ways of being in the world, water is a living entity with an active, caring role and ethical responsibilities to support plants, microorganisms, birds and animals, including people. Humans do not have the right to interfere with and prevent water from carrying out its ethical duties. Water, in turn, needs to receive care in a mutually supportive relationship with its surroundings, from refreshing rain, from fish and plants that help to keep water pure, and from humans. Yet few people understand or appreciate water in this way, often describing it as a 'resource'. The rich Welsh word *adferiad* can help us to respond in a different way. *Adferiad* [28] is impossible to translate fully, but it holds the meanings of recovery, restoration, convalescence and a return to dignity. How can we respond with care to the call of *adferiad*?

Global warming can appear as something 'out there', affecting other people and places, or as something that will impact the future. However, its effects are being felt now, throughout Hafren's course. She is flooding more and more frequently and severe floods, previously declared as 'once-in-a-century events', now occur regularly. All along her route, everyone has a flood story to share. The fruit-and-veg sellers we met in the market at Newtown told us about driving through the floodwater between Welshpool and Newtown. In 2019, the water came up to the window of their lorry. In 2022, it came up to the window of their van. On our visit to Ironbridge, the riverside Museum of the Gorge was closed due to flood damage. Global warming can seem overwhelming. What can you or I do? One can feel despair, trapped in a system controlled by others – a feeling of what the philosopher Spinoza calls *potestas*. Spinoza does, however, provide another way to respond. He proposes *potentia* where each person holds power within themselves to act in the world and be a starting point and a driver for change.

All along Hafren's course, organisations, businesses and individuals are hearing and responding to the call for *adferiad*, although so much still needs to be done at individual, local and governmental levels. The philosopher and green activist Rupert Read argues that dystopian ideas and stories of the future, which make any effort on our part seem pointless, or utopian imaginings, which comfort us with false notions that everything will be ok, are no longer sufficient.[29] He proposes that what we need now are thrutopias, which require us actively to engage together with 'how to get from here to there' and find enduring ways 'to live and love and vision and carve out a future, through pressed times'. Thrutopias encourage us to live our dreams in the present where we can, change things where we cannot, and strive together towards building a more caring world for all.

Chapter 7:
Caring
from Ironbridge, Shropshire, to Weston-under-Lizard, Shropshire

I sit on Hafren's banks at Jackfield, a mile downstream from Ironbridge. Her fast-flowing water sparkles and ripples over the rocky riverbed in the spring sunshine. Concerns over the destruction wrought by industrialisation seem far away. Calm reigns here in what was once a bustling tileworks area, Hafren busy with trows, the shouts of the boatmen merging with the noise of industry. I listen to the burr and gurgle of the water mingling with birdsong and the soft murmur of people enjoying a leisurely drink in the nearby inn garden. Grass slopes down to the water, and on the opposite bank, I can see a narrow road backed by the steep, wooded slopes of Blists Hill. This is where the last remaining trow, *Spry*, has found a dry home, standing proudly, but perhaps a little forlornly, under a hangar in the Blists Hill Victorian Town, separated from the waterways for which she was designed.

The water here looks pure – soothing to my eyes, cooling, refreshing on my warm, tired feet – but analysis in a lab would tell a different story. Run-off from fields pollutes Hafren with eroded soil, fertilisers, fungicides, herbicides and insecticides. Outlets discharge raw sewage mixed with rainwater into her during periods of high rainfall, and she has one of the highest number of incidents of pollution for any UK river.[30] Whilst this high number is partly to be expected due to her length, it is still extremely concerning. Analysis carried out at the University of Exeter for the organisation Greenpeace[31] indicates that Hafren is also polluted with microplastics – small particles less than 5 mm (0.2 in) in diameter. Fish ingest these, causing potential painful gut blockages, as well as a route for microplastics to enter the food chain. Microplastics can facilitate the spread of disease when harmful bacteria latch onto them. Also, plastics of all kinds entering rivers is a major way that plastics enter the sea. Within Shropshire, both Hafren and Worfe, which flows into Hafren at Bridgnorth, 10 miles (16 km) downstream from Ironbridge, are considered to be in poor ecological condition.[32]

Nationally, the UK does have a Water Directive Framework with a requirement for all local rivers to be classified as having good water status by 2027. Progress, however, is slow, and few rivers are on track to meet these targets. During a post-Brexit shortage of chemicals, the government in power in the UK at that time allowed water companies to pump untreated sewage into waterways. This move was seen by many as a 'green light' for water companies to carry out further pollution.[33] Changes to the legislation, which require housing developers to offset the sewage and nitrogen

Caring

run-off from their sites, mean that the government will now organise the offsetting, for example, through funding Natural England initiatives. Whilst this has potential for more 'joined-up' effective activity, there is widespread concern that funding could be 'lost' in government initiatives or in projects already announced, rather than used for genuine efforts to reduce pollution.

There are more positive government responses to issues of water pollution. In 2021, the Senedd brought in the Water Resources (Control of Agricultural Pollution) (Wales) Regulations to reduce the pollution from agricultural run-off from both severe and smaller, but ongoing, incidents. These also aim to reduce greenhouse gas emissions from agriculture and improve air quality. The Senedd argued that run-off is caused by a failure to appreciate and manage manures as a valuable source of nutrients for soil enrichment, rather than as waste products. When run-off from animal waste and nitrogen fertilisers reaches waterways it causes an overgrowth of green algae that suppresses other plants and creatures. For example, very well publicised, highly problematic pollution of this type is affecting Afon Gwy, Hafren's sister river. It is now widely established, through campaigning and the collection of citizen-science data, that the pollution is caused by the intensive chicken-rearing carried out near her and her tributaries. The Senedd has resisted significant pressure to delay or reduce the scope of the new regulations, emphasising the funding and training available to facilitate the necessary changes. Across the UK, legislation is being strengthened to give new powers to regulators to bring criminal charges against persistent lawbreakers and polluting water companies.[34] However, as campaigning organisations such as The Rivers Trust, Riverwise and the West Wales Rivers Trust point out, legislation already exists but is not enforced adequately. Only time will tell if these new measures will improve water quality. Nevertheless, these recent changes to Welsh and UK legislation show how campaigning to hold governments to account is important even where the barriers to care for of water can seem insurmountable.

Hafren

Grassroots action – individuals working together to build thrutopias at a local level – also has a critical role to play in protecting all our rivers and waterways. Many examples of this can be found along Hafren's course. Whilst this cannot solve all the problems of pollution, it can have an impact. It can also help people to develop a sense of Spinoza's *potentia* – that they, too, can do something themselves to care for their wider environment. During our visit to Shropshire, we learn about the work undertaken by organisations such as Sustainable Bridgnorth and Prevent Pointless Plastic. These organisations have successfully worked with local businesses in this small market town on Hafren's banks. Bridgnorth is popular with tourists and volunteers, keen to reduce plastic bottle use, have promoted the Refill Campaign, encouraging businesses to refill reusable bottles rather than sell bottled water. They campaign to prevent plastic waste entering Hafren due to both deliberate littering and overflowing bins, and they share information on ways to reduce plastic use in homes. Whilst in Bridgnorth, we also learn about the Severn Rivers Trust,[35] an independent environmental charity operating at every stage of Hafren's journey from her source to the sea. Their tagline is: 'It's not too late to save our rivers!' The trust works with farmers and land managers as well as volunteers from the community to engage in a variety of restoration projects ranging from citizen-science data gathering to restoring landscapes and flood-plains and creating diverse 'habitat mosaics'. Downstream from Bridgnorth, at the National Trust's estate Dudmaston Hall, volunteers are involved in conservation work with the local farmer to restore the rare and important species-rich water meadows that border Hafren here. Existing species have been identified and protected and new plants introduced. Delaying the cutting of hay until late July or early August, when the plants have finished flowering and set their seeds for next year, promotes long-term plant survival.

Across the UK, locally based wildlife trusts empower people to take action to care for and restore nature in the areas where they live. There is a wildlife trust in every county along Hafren's course.

Caring

In Powys, the area near Hafren's source, the local trust retains the historic name of the area and is called the Montgomeryshire Wildlife Trust. This trust, alone, has on average over 2,000 members and over 300 active volunteers. It also receives support from the wider community, schools and businesses.[36] Members and volunteers are involved in large-scale restoration activities, such as at the Pumlumon Project, which Peter and I visited, as well as other smaller-scale schemes. It cares for eighteen nature reserves and urban oases, such as Severn Pond Farm in Welshpool, to ensure that the county's wild creatures, including ospreys and curlews, have secure places to feed, shelter and breed. This trust also has a long-running project to protect and care for pearl-bordered fritillary butterflies, which are rare in the UK. There are only nine sites where these butterflies are known to live in Wales and six of these are in Montgomeryshire. Around Welshpool, five sites form a connected colony, known as a metapopulation.

In Shropshire, the Shropshire Wildlife Trust carries out a variety of conservation activities and campaigns to bring people closer to the wider natural world and to promote nature recovery. With the support, on average, of over 9,000 members of whom 300 are active volunteers, the trust helps to care for over forty reserves and urban and wild places.[37] For example, volunteer canoeists hold annual waterborne litter picks and, on average, remove three truckloads of plastics from Shrewsbury and two truckloads from Bridgnorth. Another scheme is the Shropshire Pine Marten Project. Pine martens had been extinct in the UK for over 100 years, so the 2015 sighting of a pine marten in Shropshire woods was an exciting moment. Since then, at least twenty pine martens, including juveniles, have been recorded on remote cameras in nine areas of the county. Pine martens are wonderful indicators of the health of woodlands. Organisations and volunteers in this area and across the UK are cooperating to learn more about pine martens and how to help them and their woodlands to flourish. The wildlife trusts further south on Hafren's journey also have conservation volunteering activities as well as special projects. In Worcestershire, volunteers help with data collection for the Worcestershire Biological Records Centre, based at Lower Smite Farm. This centre holds over one million species records, covering everything from adders to zebra spiders, and habitat data including digitised maps of county nature reserves and local wildlife sites. In Gloucestershire, trust volunteers are supporting conservation in urban and rural places. Special projects include supporting the return of pine martens to the Forest of Dean and working with the Cotswold Rivers Living Landscape Programme, which aims to reconnect and restore healthy river habitats throughout the Cotswolds.

The volunteers at all these trusts also engage in conservation work in the wider countryside, including supporting farmers who are restoring habitats. This is an important aspect of their work, especially since 70 per cent of the UK is covered by farmland. Many current farming and woodland management methods, such as the utilisation of artificial fertilisers, pesticides and fungicides and intensive grazing or over-grazing contribute significantly to river and other pollution and biodiversity loss.[38] This is not to put the blame on farmers who have been steered in certain directions by consumer demand and government subsidies that have emphasised high productivity and reduced prices. However, there are now some signs of a shift towards farming that aims to take more care of wild plants, soil microorganisms and fungi, insects, mammals and water, as we experienced at Porth Farm at Caersws. There is a plethora of terms to describe such approaches, including 'high nature value farming', 'nature friendly farming' and 'sustainable/regenerative farming'. Each has slightly different starting points and key characteristics, and the different terms tend to be used in different locations. However, all share a concern for supporting the delicate ecosystems they steward.

To learn more about regenerative farming approaches, Peter and I leave Hafren's banks at Ironbridge and travel 11 miles (17.7 km) north-east to visit the Bradford Estates office at Weston-under-Lizard.

Caring

This is the small village where my mother was the primary-school headteacher when I was a child. It is only a few days after the publication of the estate's '100 Year Vision'. Alexander Newport, heir to the current Earl, and his executive assistant, Debbie, welcome us warmly to the offices set in converted agricultural buildings. As Alexander tells us about the new vision, I can sense his excitement, his feeling of responsibility to care for the land for future generations as well as his acknowledgement of the financial privileges afforded by centuries of landownership that enable the estate to take this 'long view'. Alexander explains how, on returning to live in Shropshire in 2019 to become the managing director of Bradford Estates, he had learnt about the prediction that, if average soil degradation continues at the present rate, soils, worldwide, will only be able to support harvests for the next sixty years.[39] As the geologist David Montgomery highlights, it can take 500 years to build healthy topsoil, and less than a century to destroy it with intensive farming.[40] Alexander, therefore, decided that caring for the soil would be the foundation of this '100 Year Vision'. He has made a commitment to regenerative farming and taken 1,500 hectares (3,707 acres) of the 5,000 (12,355) owned by the estate 'back in hand' to be farmed using regenerative principles.[41]

Whilst there is no single definition, the main aims of regenerative farming go beyond a desire to conserve and care for existing ecosystems. Instead, it aims actively to regenerate those ecosystems lost through intensive farming, using methods that limit soil disturbance, increase organic matter in the soil, improve water and air quality and promote biodiversity. Key farming principles include minimal ploughing and harrowing, rotational diversity of crops, continuous crop cover and introducing livestock into arable systems. Winter cover crops feed the microorganisms and fungi necessary for healthy soil. Plants can generate sugars through photosynthesis but cannot break down minerals in the soil; soil microorganisms can break down minerals but cannot create sugars. Exchange is a perfect solution here. Soil science has now demonstrated how microorganisms can enter plant roots bringing minerals with them, exchange these for sugars, before popping back out into the soil – in what the soil scientist Dr Elaine Ingham describes as 'a never-ending rollercoaster ride'. Winter crops can be grazed by livestock that further enhance the soil organic matter with their dung, or the cover crops can be left to decompose and sink into the soil, further reducing the need for artificial fertilisers. In 2022, the estate purchased its first animals 'in a generation' – 400 Romney sheep – which will graze winter cover crops such as stubble turnips. The long-term aim is to be fertiliser self-sufficient. Cover crops also crowd out weeds, helping to reduce the need for herbicides. As food available for a variety of life within the soil increases, so do bird numbers, and the birds then eat crop-eating airborne insects. A ten-year project by the UK Centre for Ecology and Hydrology has shown that making more space for wild habitats, especially when situated on marginal areas of a farm less suited to agricultural production, does not reduce productivity.[42] With support from the Shropshire Wildlife Trust, the estate is restoring areas of wetland alongside the River Worfe, one of Hafren's tributaries in Shropshire, hoping to increase biodiversity and reduce pollution and downstream flooding.

As well as improving the soil fertility, these approaches increase the amount of carbon stored in the soil. As I had learnt when researching Hafren's boggy beginnings, carbon sequestration occurs when semi-decayed organic matter such as leaves and roots, made of the carbon that plants have drawn from the atmosphere during photosynthesis, is retained in the soil. In contrast, when soil is disturbed, this organic matter is brought up to the surface, exposed to oxygen and decays rapidly (shrivels up), releasing stored carbon back into the atmosphere.[43] Alexander cites the statistic that if soil organic matter is increased from 3 to 4 per cent, the carbon stored in that hectare will increase by 25 or even 30 tonnes.

The estate has hired a farm manager with expertise in regenerative agriculture. His first major task was to work with soil- and water-management specialists to carry out an ecological survey of the land and soils. This now acts as a baseline of soil health and was fundamental in the drawing up of a seven-year crop-rotation plan that will take incremental steps towards fully regenerative farming. The survey has identified land most suitable for crops and livestock, and other more marginal land has been designated for woodland, rewilding and habitats for natural insect predators such as ladybirds and starlings. There will be challenges. For example, this is a well-known potato-growing area, as I recall from my own childhood, and it is hard to grow this crop without soil disturbance. However, if people are going to continue to eat potatoes, not growing them just moves the problem elsewhere. The estate will experiment with ways to reduce soil disturbance when growing potatoes and include them in its seven-year rotation plan. Alexander is keen to explain to us that regenerative agriculture is not 'old-fashioned'; it embraces modern technologies, such as GPS tracking, and the estate will be working with organisations such as Innovation for Agriculture. The estate is contributing its biodiversity survey data to a University of Exeter study that compares this data with data generated by satellite images and artificial intelligence (AI) to ascertain whether biodiversity surveys can be conducted more frequently and cost-effectively, opening opportunities for more farmers to carry them out.

Given these strong arguments for regenerative approaches, we can ask what motivated the shift, especially over the second half of the twentieth century, away from more traditional farming methods, which share many features of regenerative agriculture, to the intensive agriculture widely seen today. This was, in part, a response to the increasing world population. The development of nitrogen-based fertilisers that could increase crop yields, and the breeding of animals that maximised milk and meat for human consumption, were heralded as ways to produce more and cheaper food. However, reduction in soil health, loss of topsoil, loss of wild habitats and consequent ecological loss threatens the very resilience, even survival, of the biological systems on which farming depends. Animals bred for higher yields require more costly interventions to keep them healthy. Intensive farming also requires costly borrowing to pay for initial inputs. These various factors all combine to make food production more financially vulnerable to shocks such as ever-more-frequent adverse weather. I am, of course, not a farmer or soil scientist and would encourage readers to consult books by people with such expertise and experience. They explore these issues much better than I can. What I can say, however, is that I have observed the beginnings of a shift towards many aspects of regenerative agriculture. I now frequently see winter cover-crops greening fields, boosting the soil organic matter and feeding the microorganisms that live there. In wintertime, I increasingly see flocks of sheep on arable farmland, their heads down, munching on what I now know, from Alexander, are stubble turnips.

Leaving the estate office, our heads are busy with facts and figures. It has been a lot to take in. We walk across the Shropshire landscape familiar to me from my childhood – green fields edged with red sandy soil and well-kept hedges – to a nearby copse. The bare branches of the tall oaks reach upwards above a green and brown 'skirt' created by the ivy winding around their trunks and by the lower beeches and field maples. A thick band of hawthorn and wild roses guards the edge of the wood. We squeeze carefully through this, mindful of the sharp thorns, and are quickly immersed in a multi-layered world. Our feet sink into soft leaf mulch that sits between tufts of grass, buttercup leaves and docks. We inhale the fresh air tinged with the distinctive, evocative tang of mould found in such woody places. I bend down to admire the many different greens and browns of the decaying leaves and moss-covered twigs. Nestled amongst them I see a tiny oak, a few centimetres tall, taking advantage of the light that can filter through the bare branches to the woodland floor at this time of year.

We stop to admire the aptly named lumpy bracket and turkey tail fungi clinging to tree stumps we pass. We find an oak that has blown down, its tangle of roots now exposed to our gaze and touch, and we sit down on its long, horizontal mossy trunk. The almost-

Caring

luminous, vivid green moss is soft and deep beneath my fingertips, and I suddenly wonder how it would feel on my face. I bend down and the feathery softness of the moss caresses my cheek as I breathe in its delicate fragrance. We sit a while longer. Birdsong soothes us. I can hear the melodious song of robins, treecreepers and tiny goldcrests as well as the more piercing calls of wrens and crows. The ground beneath us is damp and I think of the many layers of leaves and the webs of fungi that lie below, hidden from our view but nonetheless active, connecting and supporting the trees in an intricate, caring web of life. Webs of life are happening, too, within the red Shropshire soil surrounding this copse, where grasses and stubble turnips keep the microorganisms fed winterlong so that the amazing symbiotic relationship between them and plants can continue year on year.

Water, too, is part of this web of life. Although I cannot see Hafren from my place in these woods, I am connected to her through the water draining via the soft soil beneath my feet, entering underground aquifers, and the many brooks that criss-cross Shropshire. In turn, these brooks flow into and swell rivers including the nearby River Worfe, whose name is believed to be derived from an Old English word for wandering or meandering. The Worfe commences at Crackleybank at the confluence of the Cosford and Albrighton brooks and travels southwards for 18 miles (28.9 km) before flowing into Hafren, just north of Bridgnorth.

Sitting here, in this moment, my senses are alert to the sights, the musty smells, the sounds of the intricate, rustling, leafy world around me, and to my connection to Hafren via the intricate network of her ever-flowing tributaries. I think of the work of feminists such as Carol Gilligan, Joan Tronto, Berenice Fisher and Maria Puig de la Bellacasa who explore the ethics of care.[44] These women highlight how we are all inherently relational, responsive beings, and our human condition is one of connectedness and interdependence in a world that includes our bodies, ourselves and our environment woven together in a complex, life-sustaining web. Thinking like this opens a way to shift from abstraction and detachment towards engagement with this web and brings the potential to transform both our present and the futures we want to co-create. For these women, engaging with the sensory is not aimed at more accurately *knowing* a 'real' world, but instead at developing more *involvement* and *commitment* to it. However difficult and complex it seems, a move towards caring *in* a relationship with all those with whom we share this fragile web is possible. We *can* pause, listen, feel, touch, imagine, take small steps towards experimenting. We *can* learn and work together now towards changes at the local level, as well as campaigning for the interconnected changes needed locally, nationally and internationally. We *can* build thrutopias to carve out together, with care, a future through the challenging times we face.

Chapter 8:
Carving new routes
Weston-under-Lizard, Shropshire, and beyond

As well as the physical forces that create meanders, other special events can cause more dramatic changes in direction for rivers. These can then act as starting points for new meanderings and new learning. For Hafren, such a shock happened around 10,000 years ago when debris from melting ice at the end of Britain's last ice age blocked her route. This forced her to take a new, meandering route southwards, through what today is Shropshire, Worcestershire and Gloucestershire. Such changes in direction can be a positive thing. As Hafren tells her father in the giant's myth, she wants to take time to reach the sea: time for adventures, time to visit different places and meet different people. Like Hafren, Welsh people have also long faced situations that have caused them to leave their homes and homeland and forge new routes, new wanderings, as many Welsh stories and legends recount. In the Third Branch of the *Mabinogion*, Pryderi and Manawydan and their wives Cigfa and Rhiannon are surviving by hunting and fishing after all their flocks and herds disappeared in a mysterious mist. Tiring of this life, they decide to go to England to ply a craft. They first settle in Henffordd (Hereford) where they become successful saddle makers. Indeed, they are so successful they invoke the wrath of the local saddle makers who are losing all their trade. The two couples decide to move to another town, making shields there, and the same thing happens, so they move again and start to make shoes. Further adventures and mysterious events involve white boars, vanishing towers, golden bowls hung over fountains and a very greedy mouse.

Historical records from the 1600s onwards indicate emigration of Welsh people to the United States, India, Canada, Australia, South Africa, New Zealand and South America. This was to acquire land, find work, serve a sentence in a penal colony or found new religious and cultural communities such as Y Wladfa along the coast in the lower Chubut valley in Patagonia. This community was established in 1865 by 153 settlers who wanted to protect their language and culture from the increasing influence of England and the English language. By the end of the nineteenth century, it is thought that there were approximately 4,000 people of Welsh descent living in this area of Patagonia who attended Welsh-speaking schools and chapels. A type of Welsh, strongly influenced by Spanish, is still spoken there today.[45]

My own meandering story is bound up in more recent shocks and new directions undertaken by Welsh people who emigrated to the English Midlands between and during the world wars. The roots for this began in the developments in coal mining in the second half of the nineteenth and early twentieth century. This caused significant movements from agricultural areas of Wales such as Cardiganshire, Montgomeryshire and Merionethshire to the industrialised valleys of south Wales. The population there was also swelled by arrivals from England and beyond. Statistics from AGOR,[46] a loose partnership between the two projects Coalfields Web Materials and Llechwefan (Slateside), clearly tell this story. In 1801, the population of Glamorgan in south Wales was 70,879. By 1901, it had grown to 1,130,668. This astonishing growth was followed, after the First World War, by a period of mass emigration, caused by a decline in coal mining. Emigration was often towards towns in the English Midlands, such as Wolverhampton, and places in southern England, such as Slough, where new manufacturing industries were developing. Others went further afield to countries such as the United States, Canada and Australia. Between 1919 and 1939, approximately 500,000 people emigrated from Wales, starting new stories for these people and their descendants. The migration of young people, in search of more secure and sometimes better paid employment, continues today. This is occurring at the same time as a trend for migration of older people into Wales who, it is predicted, will become an increasing percentage of the overall population. In its 2023 Policy and Strategy document 'Wales innovates: creating a stronger, fairer, greener Wales', the Welsh Government outlines a commitment and actions to improve the employment opportunities of working-age people in Wales. These include improved infrastructure needed for homeworking and employment opportunities in the green-technologies sector.

Visiting Alexander at the Bradford Estate in Weston-under-Lizard brings Hafren's and Welsh people's stories of shocks and new directions sharply into focus for me. My mother, Beatrice, moved to teach here at Weston in the 1960s, later becoming the headteacher. Born in 1922 in the Rhymney valley, Beatrice was one of six children – four girls and two boys. Her mother was a nurse and her father worked as a coal miner and then in insurance after a mining accident. My mother told me that money was scarce but they had enough to get by. She could, however, remember some children attending school barefoot during the depression in the south Wales valleys during the 1920s and 1930s. Beatrice and two of her sisters, Morwen, born in 1917, and Elinor, born in 1928, trained as teachers in Wales and then moved to England. Morwen trained as a primary-school teacher at the Glamorgan Teacher Training College in Barry and moved to Wolverhampton. Beatrice trained at Bangor Normal College, then followed her sister to Wolverhampton where she embarked on her primary-school teaching career. She told me of the early years of teaching there during the Second World War with classes of fifty young children, very few resources and frequent air raids. She recalled an inspector saying: 'Don't worry about having enough books. Children learn to read through the pores of their skin'! Elinor was the youngest sister and she was able to benefit from the wider availability of university grants after the Second World War. She studied at Bangor University and her degree enabled her to become a secondary-school teacher of history and English. The three sisters remained close throughout their lives and had seven daughters (and no sons) between them. I was born in Wolverhampton, becoming a teacher myself in vocational (post-sixteen) education.

My mother and her sisters were not fluent Welsh-speakers, and apart from a few words and phrases such as *mam-gu* (grandma) and *nos da* (goodnight), none of their children learnt Welsh. This reflects a tragic period in the history of Cymraeg, the Welsh language, after a period of strength. Cymraeg is one of the oldest living languages in Europe, with origins possibly up to 4,000 years old. It was spoken by the Celtic Britons across the UK before the Anglo-Saxon invasion. When English sovereignty over Wales was formalised by Henry VIII's Act of Union in 1536, English became the official language of the Welsh courts and government. Wales was still, however, a Welsh-speaking country, and daily life and worship was conducted in Welsh. In 1563, the UK Parliament passed an Act for the Translation of the Scriptures into Welsh, leading to several widely used translations across Wales. Enabling people to read the Bible for themselves was important for Protestant churches and this led to initiatives for literacy education across Wales. A trailblazer in this endeavour was Griffith Jones. He was born in 1683 (or possibly 1684), in the Teifi valley, Carmarthenshire, into a 'Godly' family. He became first a shepherd and then, after much effort, an ordained minister. Around 1731, he established the 'Circulating School' movement that helped ordinary people to read the Bible in their own language. In the schools, teachers spent three months intensively teaching literacy in Welsh, before moving on to the next location. The schools were established in many different places and types of building, including the farmhouse that now lies under Clywedog reservoir near Hafren's source. Griffith was supported in this mission by Bridget Bevan, who came from a wealthy and powerful Carmarthen family. She became the movement's chief patron, providing extensive financial support and advice. After Griffith's death in 1761, she led the movement and did so very effectively. By 1773, the movement had 242 schools and 13,205 pupils. In 1854 the schools were absorbed into the National Society whose aim was to establish a school in every parish in England and Wales. Thanks to efforts such as this, it is estimated that by 1771, 200,000 people in Wales could read. This was approximately just over half the population: one of the highest literacy rates in Europe at that time.

Up until the 1850s, it is estimated that 90 per cent of the population of Wales spoke Welsh, with many in the south Wales valleys being bilingual. The nineteenth century, however, was a turbulent time in Welsh history, with popular uprisings and riots breaking out. Questions were raised in the UK Parliament about the role of the Welsh language in education. This led, amongst other things, to an enquiry headed by three non-Welsh-speaking, Anglican commissioners with no experience of Nonconformist chapels or working-class children and their living conditions. The year 1847 saw the publication of the 'Reports of the Commissioners of Inquiry into the State of Education in Wales'. This event has become widely known – one could even say it has been etched into Welsh history – as 'The Treachery of the Blue Books', named after the blue fabric covers used to bind government reports. In particular, it was the reports' comments on morality that provoked outrage. These infamous lines sum up its attitude:

> The Welsh language is a vast drawback to Wales and a manifold barrier to the moral progress and commercial prosperity of the people. It is not easy to overestimate its evil effects.

An outcome of the report was that education, which had previously been a stronghold of the nonconformist Sunday schools, became increasingly state-organised and had to be conducted in English. Speaking Welsh was banned in schools. Children caught speaking their native tongue were humiliated by having to wear a wooden board around their necks and sometimes they were also physically punished. In the industrialised valleys of south Wales, the influx of workers from England and beyond also weakened the speaking of Welsh and the language became increasingly marginalised. I recall my mother saying that her own parents spoke English and Welsh but only spoke Welsh at home when they did not want the rest of the family to understand. My mother and her sisters spoke only English at school and at home. As well as this practice being encouraged by the state, my mother's parents also saw English as the language to gain an education and the skills needed to leave the valleys to find work and a healthier life outside Wales.

By the mid-twentieth century, these policies and beliefs had become so widespread that it was feared that the Welsh language could die out. This led to a strong revival movement, including the setting up of Cymdeithas yr Iaith Gymraeg (The Welsh Language Society) in 1962. In 1993, 457 years after Henry VIII's bill banning Welsh from public life, the Welsh Language Act gave Welsh equal status as a language of the UK. All signage and public documents were to be produced in both languages. Now all children in Wales, including my sister Rachel's children, study Welsh up to the age of sixteen. There are also schools where all subjects are taught in Welsh, and there is now easy access to television broadcast in Welsh. Like many adults who have roots in Wales, I have started a modest study of the language. In the Welsh Government's 2023 Annual Population Survey,[47] 29.7 per cent of the population (906,800 people) aged three or older were able to speak Welsh, with 15 per cent reporting that they spoke Welsh daily. Of additional significance is that it was the five-to-fifteen age group that was more likely to report that they could speak Welsh (50.7 per cent, 248,100 people). The loss of language and the accompanying loss of cultural identity and historical, musical and literary riches that a language carries, is beginning a slow process of reversal.

We had made this journey today, from Hafren's banks at Ironbridge to Weston-under-Lizard, to hear about the estate's plans for regenerative agriculture, but we had also come to visit the former village school, now a foster carer hub and training centre. Although my sister and I had lived in the neighbouring village and attended school there, we had also spent a lot of time in this school and still feel proud of my mother's role as headteacher. As Peter and I cross the busy A5 that runs through this villlage, I feel quite nervous. It is a long time since I have been here and it carries many memories. We pass red-brick gabled cottages with neat front gardens surrounded by picket fences. These were originally built in the nineteenth century to create a 'model village' with the school as its centrepiece. The high wall which used to enclose the school and playground has been removed and I can immediately see the Victorian building. Time rolls back as I recall the movement and shouts of the children who used to play in front of the school's high windows, pitched roof and tall chimneys. These architectural features are typical of schools built in response to the 1870 Elementary Education Act in England and Wales that established a system of compulsory education managed by 'school boards' – locally elected bodies that drew their funding from local rates.

Hafren

Rachael, who works at the hub, welcomes us. With amazing warmth and enthusiasm in her voice, she tells us about the activities now carried out here. Her love and care for the children and families that she supports shines through, a silver thread extending back through time to the love and care my mother, the other teachers, the cook and the school cleaner here had given their pupils. My mother would certainly have enjoyed meeting Rachael and knowing about the work that continues here. The building itself had changed very little: two high-ceilinged classrooms (infant and junior), toilets and a narrow corridor leading to a small kitchen and what was once a tiny office. My mother taught the junior class as well as carrying out the responsibilities of being head. At times she taught the whole school together, especially as numbers dwindled. As we look around, I can picture in my mind all the wall displays, the fêtes, the Christmas parties. Looking down the narrow corridor towards the kitchen, a memory comes to me of kneeling there, sorting through fabric scraps stored in a box to find pieces suitable for a craft project, whilst my mother catches up with paperwork in her office. Young enough to still believe in Father Christmas, I had been very surprised to find a piece of the black-and-white polka-dot fabric used in a doll's outfit that I had found in my Christmas stocking a few days earlier. Standing here today I feel I better understand and appreciate the long apprenticeship in teaching my mother gave me: the sharing of her skills in preparation, imagination, communication, creativity, perseverance and kindness. This apprenticeship has been an important tributary in my own growth as a teacher in further and higher education and has influenced the route I have carved through life. It has evolved and flowed onwards too into James's life as he shapes his own route as a workplace trainer, learning from us both the importance of allowing space for creative responses. Another thing he says he learnt from us is the importance of taking teaching seriously but not taking oneself too seriously.

In the infant classroom, with its vaulted ceiling supported by metal fretwork, I picture the busy and creative life of both the foster centre and the school. I ponder the value of meandering in education. A few months prior to this visit I had sent for copies of some early records from the school that are now kept at the National Archives. On reading these I had been struck by how the tone and language of government inspectors' reports written today have hardly changed from these early examples. In a 1930s report, an inspector praises the school for children's skills and knowledge in reading, writing and composition and their attention to the tasks set by the teacher, which they also maintain when the teacher leaves the room. There *is* a need to encourage focused attention to develop the skills and knowledge associated with schooling. However, it is also important to ask, 'Is this all that education can or should be?' This is especially relevant in this era with its ecological and climate problems that do not have obvious solutions. As research at the University of California, Berkeley, indicates (see Chapter 6), much of young children's thinking-time is spent meandering, which then helps to encourage creative ideas – carving new paths, just like rivers. In addition to focused time, students need opportunities to think and explore freely, unfettered by the existing rules of knowledge. They need time to be creative, imaginative and speculative. Mind-wandering, playfulness and allowing thoughts and feelings to meander has an important role if creativity and originality are to be part of education. The philosopher Hannah Arendt has a wonderful way of expressing this, arguing that education is where:

> we decide whether we love our children [and young people] enough not to expel them from our world and leave them to their own devices, nor to strike from their hands their chances of undertaking something new, something unforeseen by us.[48]

My mother knew this well. The setting of the school supported her with access to a country park and estate on her doorstep, providing a place for the children to explore and connect with the teeming life outside the four walls of a building. The UK-wide campaign 'Plant a Tree in '73'[49] was a wonderful opportunity for the children (myself and my sister included) to learn about, experience and engage with the trees and wider natural world around them. This included moments of playfulness in the forest adventure playground with its swaying rope walks between high branches, and a zip wire to enjoy a feeling of flying amongst the trees, the wind touching our cheeks. The wider park also provided moments for dipping our feet in the local streams, kicking up leaves, searching for conkers and acorns and for experiencing the scents and sounds all around us. The work of my mother's tiny school was recognised in a national competition set up to encourage schools to plant and to learn about trees in their local environment. Weston was one of the twenty winning schools whose project work was displayed in Trafalgar Square in London. I still treasure the heavy bronze disc, about 10 cm (4 in) in diameter, decorated with a beautiful arching tree, awarded to each winning school. Engraved on the back are these words:

For good work in
TREE PLANTING YEAR
1973
Presented by the
SECRETARY OF STATE FOR THE ENVIRONMENT

Whilst most schools are not set in country parks, spaces for playfulness and exploration can still be found – the corner of a playground where plants push through the tarmac or grow in a raised bed or in a nearby urban park or woodland. The UK Government is introducing a new GCSE in natural history. As the environmentalist and campaigner Mary Colwell comments, this GCSE will:

reconnect our young people with the natural world around them. Not just because it's fascinating, not just because it has benefits for mental health, but because we'll need these young people to create a world we can all live in, a vibrant and healthy planet. [50]

Exploring the archives from the school also draws my attention to a caring role for water that perhaps many (but not all) people take for granted. In 1909, the school failed its inspection and was deemed 'unsatisfactory' because of the state of the outdoor earth closets that had no water connected to them. It was claimed that the stench posed a health risk to the children in the adjacent classroom. The school records contain a lengthy correspondence between the estate that owned the school property and the National School Board. I had not really thought before about the importance of the letter 'W' in the abbreviation W. C. for 'water closet'. I do recall, however, that as a young child in the 1960s, I did not much like the indoor toilets, which were subsequently built in the entrance porch of the school, finding the smell of disinfectant on entering the building unpleasant. I also remember my mother gently rebuking me for mentioning this, saying that the school was lucky and proud to have these indoor toilets with running water for flushing and handwashing. As I read these records, I feel Hafren and the web of water running through Shropshire nudging me, reminding me of the importance of water and its role in all our lives. It is so easy for those who have easy access to clean, running water to take it for granted. We need to care for water, and in turn, let water care for us.

Chapter 9:
Travelling onwards
from Upton Warren, Worcestershire, to Deerhurst, Gloucestershire

Peter and I wait until May to undertake the next part of our journey following Hafren's southerly course through England. Our destination is the Christopher Cadbury Wetland Reserve at Upton Warren near Droitwich in Worcestershire, about 40 miles (64 km) downstream from the school at Weston-under-Lizard. We are going to visit Hen Brook, part of the watery web that builds Hafren's powerful flow through Worcestershire. Hen Brook flows into the River Salwarpe, itself created by the confluence of the Battlefield and Spadesbourne brooks at Bromsgrove. The Salwarpe flows for 20 miles (32.2 km) before joining Hafren, just below Droitwich.

Travelling onwards

At Hen Brook, the Worcestershire Wildlife Trust has created a rewiggling project. 'Rewiggling' is a term used to describe the removal of raised and reinforced banks built previously to contain and straighten water-courses. It allows water to overflow the banks and restores meanders. These act to retain water and help reduce flooding in the network of rivers further downstream. They also create more diverse habitats to support a wider range of flora and fauna. At Hen Brook, restoring the meanders has created marginal, watery zones for plants such as water mint and pink water speedwell. As well as being valuable in their own right, these plants provide cover for fish and dragonflies and their nymphs (larvae). The next stage of the project is the dredging of Hen Pool, which has become dominated by common rushes over time, and the creation of a new scrape (shallow indentation to create a wetland zone) and a swampy area. These will provide a tranquil zone for water voles, demoiselle dragonflies and shy birds such as green sandpipers, water rail and snipe.

The late-spring day is overcast but mild, holding the promise of warmer days to come as we wander beside the brook. We absorb its gentle, easily overlooked beauty and the peace it shares with us. The muddy banks are dotted with ivy-clad trees rising above areas of low-lying green vegetation. May blossom is in full flower, frothy, extravagant on the dark branches of blackthorn. We can smell the mud – rich, earthy – as it squelches under our feet. The call of crows and pigeons blends with the burbling high notes of the water that runs in rivulets through the mud before joining the brook. Shy moorhens, with their distinctive red beaks, hide behind the reeds as orange-tip butterflies dance by and beige-bottomed bees investigate the blossom. We enter a bird observation hide next to a pool edged with tall reeds. A pair of swans are nesting here, and we arrive just as they change places on the large nest. This is made from a mass of twigs, the spaces between them packed with mud. As the male swan leaves the nest, the female clambers on and turns each egg carefully. She preens her breast, which removes some of

Hafren

the feathers, helping heat transfer from her body to the eggs, then she settles down. She does not rest, however, but continues to build and maintain the nest, stretching her neck out to reach more twigs and mud to pack into the existing structure. I feel privileged to share this caring, intimate moment from within the stillness and darkness of the hide.

Walking beyond the brook and wooded area, we arrive at a more open space of hawthorn hedges and rough grass sloping down to one of the pools of the reserve. Unusually, this is an inland saltwater pool due to the underground natural brine (salt) springs that have existed in this area for millennia. Entering a hide beside the lake, we can see and hear a wide variety of birds. Some are more usually associated with estuaries, such as oystercatchers with their reddish-pink legs and sword-like, carrot-coloured beaks, designed for digging into deep sand. Avocets, with their black-and-white plumage, sweep their long, slim, upward-curving beaks through the shallow water in search of aquatic insects, worms and crustaceans. These birds became extinct in the UK in the late nineteenth century but began to re-establish themselves as a British breeding species in the 1940s. This was helped by the closure and flooding of beaches on the east coast, done as a defensive measure in the Second World War. Whilst still scarce and protected in the UK, avocet numbers have grown over the last seventy years.

As we look around the lake we see majestic lapwings, with their distinctive black head-plume, strutting across the mudflats and small, slender Baird's sandpipers dunking their short beaks at the water's edge, foraging for invertebrates. Black-headed gulls circle and land, their dark heads and red beaks and legs making identification easy. Amongst the busy, noisy birdlife we spot a black-tailed godwit. This large, elegant migrating bird has long legs and distinctive black markings on its wings, and, of course, on its tail. Rare in the UK, they are classified as highly endangered (red listed) and protected. Encouragingly, godwit numbers are now increasing in Iceland, where they spend the summer months, due to changes

Travelling onwards

in agricultural practices there. We are very new to birdwatching and are delighted to meet Pat and her husband who help us by sharing their bird knowledge. Being with the birds gives them tremendous pleasure, and Pat tells us of her particular delight in the majestic dance of lapwings. When they ask us why we are here today, we tell them about our journey with Hafren and how we have come to see the rewiggling project. Pat tells us about their local river, the Arrow, which rises in the Lickey Hills and travels broadly south-eastwards across Worcestershire before flowing into the River Avon near Salford Priors in Warwickshire. The Avon, in turn, flows into Hafren at Tewkesbury: another interconnected watery web. One of their favourite spots is a particularly meandering section of the Arrow, near their home. They have observed how the water's flow does indeed slow down as it passes through this winding section.

It is a short train journey of a little over ten minutes from Droitwich (the nearest station to the nature reserve) to the city of Worcester where the cathedral is perched high above the riverbank to keep it safe from Hafren's flooding. I am travelling there today to visit Diglis Island and its new fish pass, a series of shallow 'steps' constructed beside a weir to enable fish to swim upstream without having to make big jumps that can injure them and consume considerable energy. This new fish pass, constructed by the Unlocking the Severn project, is one of several built to restore connectivity for fish along Hafren's course. It is about half a mile downstream from the town centre, and as I walk along the paved riverbank, I am passed by joggers and cyclists as well as people strolling, chatting with friends, takeaway coffee in hand, enjoying time beside the flowing water. I pause by the cathedral, which can be reached from the bank here via steps accessed through the Watergate Arch. This arch dates from 1378, built to provide access from Hafren to the cathedral's monastic precinct. I study the high-water records on the wall beside the gate. A brass plaque tells me that the heights of significant floods have been recorded here since 1672 and that the highest flood occurred in November 1770.

Hafren

I estimate that the plaque, set at the height of this flood, sits almost 5.1 m (17 ft) above the current level of Hafren's bank. The dates and heights of other floods are carved into individual stones set into the wall at the height the floodwater reached. The floods in 1886, 1947 and 2007 almost match the record for 1770. There are so many stones marking significant floods across the centuries, and these seem to be happening more frequently. Each one of these floods was significant for those living and coming here. I look across to the Chapter Meadows on the opposite bank, which have long served as a flood-plain for Hafren in the heart of the city. I take a moment to remember seeing them filled with water in the floods of 2007 when I visited the city for the funeral of a dear friend. Today, the cries of children playing on the meadows mingle with the calls of seagulls and the squabbling of magpies. Swans glide by, numerous here because of the daily feeding on this stretch of Hafren by The Swan Food Project, set up because the heavy burden of silt carried by Hafren limits the quantity of plants in the water for them to eat.

I soon arrive at Diglis Island. The project here was inspired by one of the UK's rarest fish, the twaite shad, a member of the herring family. Hundreds of thousands of shad used to migrate up the river each year to reach their natural spawning grounds, such as those found all the way up-river near Welshpool. Smaller than salmon, they prefer the finer shale found there as they can move it more easily with their tail fins as they prepare the riverbed for their eggs. Records show how shad were often preserved with Droitwich salt and were once so numerous that excess catches were even spread on fields as fertilisers. However, the building of the weir and lock at Diglis in the mid-nineteenth century blocked shad migration upstream. Now, the Unlocking the Severn project has provided fish passes on six barriers on Hafren and the River Teme, restoring access for the shad to 158 miles (254.3 km) of river habitat. This has also benefitted other fish species including salmon and eel. DNA testing on the fishes' discarded scales and cells found in water samples from the river show that shad have now navigated past

Travelling onwards

all four barriers on Hafren and begun to reclaim their spawning habitats for the first time in almost two centuries. The DNA testing also provides information on fish numbers, and early results indicate these are increasing. Another important aim of the project is to reconnect people with Hafren, and the project works with schoolchildren and community groups on a range of educational and wildlife and heritage initiatives. Volunteering opportunities and citizen-science projects are key to understanding fish migration and other behaviour patterns.

I cross Hafren using the pedestrian bridge just south of Diglis Island and walk a short distance back upstream to the fish pass: an engineering wonder of concrete and rushing water. The construction here includes a special feature – a viewing window that allows us to see the fish swimming beneath Hafren's surface. May is the month of the shad migration upstream. I am enthusiastically welcomed by Patricia and Graeme, who, along with other volunteers, facilitate visits to the fish pass on the open days advertised on the project's website. Feeling excited, I wait for my allocated time slot and am soon descending the steps into a concrete 'bunker' where a viewing window is set into the wall facing the base of the pass. This is back-lit with infrared bulbs that cast an orangey-red glow over the water. Volunteers Rashi, Heather and Penny provide some background information about the project as we wait. Soon our patience is rewarded, and I have my first underwater sighting – two bream, deep-bodied, bronze-coloured fish with high, curving backbones. It is a wonderful moment. Previously, I could imagine that fish were there, but now I feel so privileged to see them for real. Soon, more bream swim by and then a shoal of tiny silvery bleak, their distinctive eyes, large compared to their small size, reminding me of the shoals I associate more with tropical corals.

My first twenty-minute underground session ends, but I am invited to wait at ground level for a lull in visitor numbers. This gives me the opportunity to talk to Jamie, a scientist at the University of Hull, who is here today for an interview with a scientific journal.

Hafren

He explains to me the various projects to count and tag shad and how the DNA testing of discarded fish scales works. On my second visit to the observation window, no fish swim past, but on my third visit, I see so many – two mighty salmon, a roach with its distinctive reddish fins and tail, two small silvery dace and several more bream. I am amazed at the size of salmon – almost 1 m (3 ft) long. Having the opportunity to see all the different fish passing through the same window really gives a sense of scale. As well as helping shad, the fish pass facilitates the movement of all the fish I am seeing today, as well as others such as eels and lampreys.

Then, there is great excitement: a twaite shad swims past. It is a wonderful moment, seeing this small silver fish with its distinctive forked tail make its way past the glass then disappear on its onward journey upstream. Heather calls out: 'Who'd like an "I've seen a twaite shad" sticker?' Completely forgetting that I am an adult, I raise my hand and energetically call out: 'Me'. Heather laughs and hands me a sticker. The shad sighting was carefully added to the tally recorded on a board. Within a few minutes, two more shad dart very quickly past. I feel so lucky to see three of these rare fish. There is a sense of celebration in the small chamber, and Heather tells us how they are known as pioneer fish, setting off to explore. Although shad have not been able to swim upstream since the four weirs between Worcester and Stourport were built between 1843 and 1845, these fish are still driven by some kind of 'memory' to migrate upstream from the small breeding site shad have established just downstream from Worcester. Jamie, the scientist, has now joined us in the underground chamber, and he tells us that part of the scientific project at his university involves sending data from the fish numbers and migration patterns to a scientific project in Amsterdam, where information from across the globe is being analysed to try to understand this and similar unknowns. There is so much we do not yet know about animal intelligence. I ask Jamie what has caused the decline in shad numbers. He explains they have been negatively impacted by climate change, warming

Travelling onwards

and rising sea levels, pollution and barriers to migration. Increasing numbers of smolt (young fish) are being eaten as they travel out to sea due to less food being available for other fish species, leading to further falls in fish numbers in a vicious cycle. Projects such as Unlocking the Severn that increase the number and suitability of shad breeding sites are important opportunities to help reverse this trend as well as providing opportunities to understand more about fish migration. It has been a wonderful afternoon.

Travelling home from Worcester, the train approximately following Hafren's journey downstream, I can still feel the excitement, pleasure and the special privilege of having seen so many fish: this life in Hafren's waters, normally hidden from view. It has also been an amazing opportunity to learn more about them from the enthusiastic scientist and volunteers. As I travel south, I think of a trip that Peter and I made together the previous autumn to Upton Ham Nature Reserve, at Upton-upon-Severn, approximately 9 miles (14 km) downstream from Worcester. 'Ham' is the Anglo-Saxon word for meadows/water meadows, and these were often looked after using traditional methods governed by the calendar. Upton is a pretty, historic town, with black-and-white-timbered buildings nestling beside Hafren. The bridge is a well-known, strategic crossing point, fought over, for example, at the Battle of Upton in the seventeenth century during the English Civil War.

On the chilly autumn day we had made our visit, clouds skittered across an overcast sky, and Peter and I had been glad to pull our warm layers more closely around us. Leaving the town behind, we were soon making good speed beside the fast-flowing water fringed with willows, alders and poplars. Sheep were calmly cropping the grass and the silhouettes of occasional oaks punctuated the flat ground. In the distance we could see the striking outline of the Malvern Hills. The seasonally flooded meadows here at Upton Ham are still tended using the traditional principles named after the religious festival of Lammas. Nature-friendly and regenerative farming are increasingly taking note of traditional patterns and practices such as these, which create ideal conditions for myriad plants, insects, birds and mammals. Hay is cut in late July, which allows wild plants to flower and seed, and ground-nesting birds to fledge their young before the haymaking. Sheep are then introduced for a limited period of grazing, between Lammas, which falls on 1 August, and New Year's Day. Regular flooding spreads rich river sediment over the pasture. Combined with the sheep dung, the sediment boosts the soil fertility naturally, avoiding the use of artificial fertilisers. Since flooding further limits the autumn and winter grazing, the plants are not cropped too closely. The wide variety of plants on these water meadows store the carbon that they have drawn down from the atmosphere in their dense mesh of closely packed, and often deep, root structures. The deeper soils, which build up on flood-plains due to the sediment deposited over thousands of years, create a good distribution of carbon throughout the whole soil profile, not just in the very top layer.

Lammas meadows were common land, where local commoners had grazing rights. Strictly speaking, these ones here at Upton are not 'true' Lammas meadows, as they were enclosed in the late nineteenth century and are no longer common land. Today, parts of the water meadows are owned and cared for by a local farmer, with the support of the conservation charity Plantlife; other areas are owned by the Kemerton Conservation Trust. However, they are still stewarded using Lammas traditional methods and timings. The area has been designated a Site of Special Scientific Interest (SSSI).[51] It is home to meadow foxtail, mousetail, meadow vetchling, meadow saffron and great burnet, a plant particularly characteristic of this type of meadow. The nationally scarce water dropwort is also found here. The water meadows provide a breeding habitat for curlew and redshank, and in winter a large number of snipe feed on the wet grassland. On our autumn walk we had been too late to see flowers but did see many grassland fungi, indicating the hidden, vital life in the soil beneath our feet. The parts of a fungus above ground are its fruiting body that produces and releases its spores.

Travelling onwards

The living body of the fungi is formed from mycelium, which is made of a web of filaments called hyphae. This is usually hidden in the soil. In ways that are not yet fully understood, mycelia form relationships with soil microbial communities and increase plant nitrogen acquisition. The web of fine filaments also improves soil stability and helps the soil retain moisture. Soil disturbance and the use of fertilisers damage these important webs that are growing beneath the water meadows. Their role is becoming better known, beginning to match the interest shown in the more widely discussed webs under our woodlands. Lammas practices, with their limited soil disturbance and avoidance of artificial fertilisers, help to protect these important networks. The cropping (but not overcropping) of the grass created by sheep also supports the fungi on the meadows: if grass is left too long, fungi do not waste their energy producing their fruit, as the wind will not be able to disperse their spores and increase their numbers.

Lammas Day was an important date for both the Celts and the Anglo-Saxons, and in the Christian tradition the word 'Lammas' is derived from the Anglo-Saxon *hlaffmaesse*, which means 'loaf mass'. It celebrated the first fruits of the harvest, and a loaf was brought into the church. Lammas, called Lughnasad or Lughnassadh, is also a 'cross quarter' day in the traditional Wiccan 'wheel of the year'. It lies half way between the summer solstice and the autumn equinox and marks the start of the harvest season, particularly the grain harvest. In the Irish Celtic tradition, the festival is called Lughnassadh and is dedicated to the sun deity Lugh, who infuses wheat with his power and is sacrificed when it is harvested.

In the Welsh tradition, the festival is known as Gŵyl Awst and the solar deity at the heart of the story is widely understood to be Lleu Llaw Gyffes. When thinking about bordering and crossing earlier on my journey, I had learnt from the Fourth Branch of the *Mabinogion* that, since Lleu could not marry a human, his uncle Gwydion and King Math had conjured a woman from flowers to be his bride, calling her Blodeuwedd (flower face). Thus, the verdant spirit of spring was trapped in a woman's form. Angharad Wynne[52] suggests this can be interpreted as a metaphor for the sun god Lleu 'husbanding' the spirit of burgeoning Nature. However, Nature cannot be tamed for long. When Lleu is away, Gronw Pebr, the wild lord of a neighbouring kingdom, rides by with a hunting party. Blodeu(w)edd and Gronw feast and sleep together for three days, and before Gronw departs, they hatch a complex plan to kill Lleu. The ensuing adventures and events make a marvellous myth, and one that makes good sense within the patterns of the year. Nature needs to be released from her union with the sun (summer) to enter into partnership with the hunter who rules autumn and winter.

Hafren travels onwards from Upton, passing through Tewkesbury, where her waters are swelled by the River Avon. There are several rivers in the UK called Avon so this one in Tewkesbury is sometimes called the Warwickshire Avon as that is the county where it arises. The word 'avon' shares the same root as the Welsh word *afon* (river). *Afon* is also pronounced 'avon', since a single 'f' in Welsh is pronounced as 'v'. Both are derived from the word *abona*, the Common Brythonic word. In the Cornish language, the word for river is also *avon*, and in Breton, the word is *aven*. The shared words emphasise the Celtic links between these different nations that were pushed back into separate parts of Britain by the Anglo-Saxons. Building the town of Tewskesbury beside not one, but two, major rivers was a brave move, but still considered advantageous when trade relied on river transport – the situation for most of human history. The area around the town is subject to flooding and several areas are still reserved as 'ham' (water meadows) to encourage flooding there and ease the pressure on the town centre. This town does, however, still experience regular floods, the most severe mirroring those in Worcester. Despite this, as in Shrewsbury, residents here have adapted and found ways to enjoy living beside these, often tempestuous, flows of water.

Two miles (3.2 km) south-west of Tewkesbury, Hafren reaches Deerhurst, a small village I was keen to visit after researching these

Travelling onwards

contested borderlands between England and Wales. So, on a very wet and windy early summer's day, Peter and I set off to visit the area. We start at the Coalhouse Inn, about a mile downstream from the village. It dates from the sixteenth century and was originally called the White Lion. It became known as the Coalhouse Inn due to its popularity with the men on coal barges using the now-abandoned 2-mile- (3.2-km)-long Coombe Hill Canal between Hafren and the old coal wharf at Coombe Hill. A pause in the rain enables us to sit outside for food and coffee. We look across a flat, wide stretch of grass that extends towards Hafren's fast-flowing water, which we can glimpse through gaps in the thick band of trees that line her banks here. This stretch of Hafren has some of the highest water discharge of UK rivers with an average 107 m^3 per second (3,779 ft^3 per second) passing through the hydrometric measurement station at nearby Haw Bridge (the volume of water flowing through a river channel is measured this way). It is hard to comprehend these numbers in the abstract, so I try to picture 107 tea chests full of water, each one measuring 1 m × 1 m × 1 m (3 ft × 3 ft × 3 ft) passing the measurement device every second: a huge volume of water.

We fall into conversation with the pub's landlady and Dave, one of her regular local customers. I ask them what it is like to live so close to Hafren, especially in an area well-known for flooding. 'It is hard,' replies the landlady, describing how in the 2019 floods the water came up to the window ledges of the ground floor. She adds that, rather than floodwater coming directly from Hafren overspilling her banks in front of the pub, the water had come in rapidly from the low-lying, flat field behind the pub that had filled with water from flooding further up-river. As I expected to hear after my recent visit to Worcester, the floods in 2007 were even more severe and the water came up to the windows of the first floor. There was also, however, a sense of pleasure from living alongside Hafren – the peace, the ever-changing views, the walks along her banks. Dave described his boat and the enjoyment this had given him and his family, as well

Hafren

as other water-based activities they had enjoyed that even included waterskiing. This reminds me of our visit to Shrewsbury and the pleasure the people there also experience from living alongside Hafren, ranging from active sports to quiet sitting: this aspect of enjoyment is sometimes lost in all the talk of flooding.

Leaving them to their reminiscences – we have noticed that everyone living near Hafren has flood stories they love to share – we set off to walk the short distance upstream to Deerhurst. The dense screen of willows, alders and beeches that divide the green fields from Hafren's banks occasionally gives way to small openings where we can scramble down to her grey-brown water and watch as small branches are swept along. The rain returns and, as it intensifies, we feel increasingly part of an all-enveloping watery world. The summer warmth makes this surprisingly pleasurable. We enjoy the patter of rain on our waterproofs, the squelching sound of our boots as they sink into the muddy ground, the feel of water dripping from our hoods onto our faces and the clean taste of the water drops.

We see regular signs saying, 'Private Fishing, Birmingham Angling Society', but our access to the bank is not blocked thanks to the public footpaths along Hafren's lower reaches. We do, however, see a sign on a gate stating: 'No Entry. Private Elver Fishing.' I am confused by this as I know that elvers, the name given to young eels swimming upstream to spend time in fresh water, are internationally listed as critically endangered. They have an incredible life cycle greatly threatened by climate change and the impacts this is having on sea levels, sea temperatures and ocean currents. After many years in European river systems, mature eels swim back to their breeding grounds in the Sargasso Sea, in the Western Atlantic. The eggs they lay there are then wafted on the ocean current towards Europe, hatching during this journey into 5-mm- (0.2-in)-long larvae called leptocephali that continue to drift eastwards. Their 4,000-mile (6,500-km) journey takes between one and two years. Leptocephali mature first into transparent tiny eels called 'glass eels', then into elvers. I discover that the government

Travelling onwards

patrols and monitors Hafren's banks here between February and May, the season for the elver migration, and permit a very limited amount of licensed fishing using traditional elver dip nets. In total, around 300 individuals are allowed to catch elvers in the lower stretches of Hafren and Gwy.

After passing through a small woodland, the footpath opens into a wide, flat, grassy area, and we can see Deerhurst about 400 m (440 yd) from Hafren's banks. We follow a diagonal path towards the first row of low cottages, keeping a watchful eye on the cows sheltering under a large oak standing proud in the open space. Fortunately, the cows seem more interested in staying dry than in us. We pass another oak with a huge hole formed within its trunk, large enough for me to get inside easily. It makes a lovely den, reminding me of childhood days spent along the lanes not so very far from here. We soon reach the cottages and the gate leading to Odda's Chapel. Built in 1056 by Earl Odda in memory of his brother, it is one of the most complete Anglo-Saxon chapels still standing today in England. In the seventeenth century it was incorporated into the neighbouring farmhouse to extend the kitchen and provide an additional bedroom. It was rediscovered in 1865 by the Revd George Butterworth, who was following up clues in local records about the location of a chapel here. The small stone building is still attached to its black-and-white-timbered neighbour.

As we enter through an archway into two distinct areas – the nave and the chancel – we are struck by the peace. The only sound is the muffled patter of rain. The large, rough-hewn blocks are mossy-green in places, the colours merging into the yellowy-cream of the drier stones. Wooden beams support the steeply pitched, tiled roof. I place my hand on a stone, feeling its damp chill and – somehow – a connection with those who have been here before during this place's varied history. A ledge runs around part of the wall, and we sit quietly for a while, enjoying the calm. Walking on, we follow a path that joins a narrow road towards the Priory Church of St Mary. As we approach the main part of the village, we pass through a wide, green, metal floodgate, over 2 m (6 ft) high, topped with black-and-yellow 'alert' banding. The gate is part of the extensive flood defences in the village. This, as well as other similar gates placed strategically around the village, closes in periods when flooding is imminent. There are also permanent levees (floodbanks) that were improved after the severe flooding of the village in 2007. How long, however, will these defences remain effective? According to maps created by Climate Central[53] – an independent not-for-profit group of scientists and communicators – Hafren will be permanently lapping at these gates by 2050 if global warming causes sea levels to rise as currently predicted.[54] We follow the lane and take the short path through the churchyard to enter the large square-towered church, where we learn that a religious foundation was established here in 804 CE, when Æthelric bequeathed extensive lands to the community. Records indicate that in the first half of the ninth century, Deerhurst was one of the most important religious foundations in the Anglo-Saxon kingdom of the Hwicce, a sub-kingdom of Mercia. It was here in the second half of the tenth century that St Alphege began his ecclesiastical career. He is known today as the patron saint of kidnap victims due to his own experience, when he was Archbishop of Canterbury, of being captured by Viking raiders in 1011.

I am drawn to the softly lit, Anglo-Saxon font: a simple bowl carved from a single block of limestone set on a stem of the same width. Believed to date from the ninth century, it was at some stage removed from the church and possibly used as a farmyard water trough or garden ornament as well as spending time in another church nearby. It was only reinstated here in the middle of the nineteenth century. The bowl is totally covered in spirals linked into pairs by strong, straight lines. A frieze of leaves, buds and bunches of berries runs around the top and bottom edges of the bowl. The stem is also intensely decorated with more linked spirals, ribbon-bodied creatures and leaves. I am fascinated by this font. I learnt very little about the Anglo-Saxons at school, told that once

Travelling onwards

the Romans left Britain in the fifth century the country entered 'the Dark Ages'. Whilst I was told this referred to the lack of information from written accounts, there was also an implication that this was an 'uncivilised' period of history. Looking at this font, so expertly and effectively decorated, I question this portrayal.

We make our way back through the village and the broad field that acts as a flood-plain for Hafren. We squeeze through the alders and willows that line the bank to stand beside the grey, fast-flowing water. The bank is sodden, and it is far too wet to sit so we retreat a little into the trees and lean against a larger alder. Two swans glide by, travelling upstream against the flow of the water rippled by rainfall and a light wind. I wonder which fish might be swimming unseen by us, under the surface, and treasure the memory of my afternoon in Worcester at the fish pass that afforded me a glimpse into this watery world. I feel a sense of gratitude to Hafren for bringing us to this place today, filled with history, peace and beauty. I picture the font I have just seen with its detailed carving and reflect on the skill of those who made it: a skilfulness that challenges my preconceptions about the Anglo-Saxon period. I think, too, of the traditions and stories of Lammas and the Lammas meadows at Upton. The value of these traditions as well as new insights about the Anglo-Saxons and their lives are coming to the fore. As I stand here beside Hafren's ever-flowing water, I wonder what we could learn today if we were to try to understand with a little more openness and humility the ways of life of people who have come here before us.

Chapter 10:
Feeling the pull of the tide, the pull of home
from Gloucester, Gloucestershire, to Oldbury-on-Severn, South Gloucestershire

After Deerhurst, Hafren begins to turn in a south-westerly direction. Increasingly influenced by the tides, there is a feeling that she is getting ever closer to coming home to Wales at Chepstow, to reuniting there with sister Gwy and returning with her, in silvery splendour, to the sea. It is the sea that is, in turn, the starting point of the rainwater that creates their boggy beginnings high in Pumlumon and takes them back to their father, the giant. At one time, the effects of the tide reached as far as Worcester, but these days, weirs and locks hold back daily tides, except for some particularly high spring tides (spring tides, somewhat confusingly, occur twice a month around the time of new and full moons). These tides can overtop the weir at Gloucester and make Hafren tidal as far as Upper Lode Lock near Tewkesbury.

It is a short journey of 10 miles (16 km) from Deerhurst to the ancient settlement of Gloucester. Unlike in Worcester, where Hafren flows through the centre of the city, here in Gloucester she passes to the west. The archaeological and geological records of Hafren's historical and present-day routes are not straightforward, but it is generally agreed that several millennia ago there was a single meandering channel here, shaped like a sideways, gently curving

Hafren

letter 'U', with the closed side of the 'Ↄ' east of Hafren's present day course. A bifurcation then occurred. This is a natural part of a river's meandering processes that occurs when water overflows the outer bank at the start of the 'Ↄ', floods downwards and re-joins the original course at the bottom of the loop, forming two channels with an island in the middle. The straighter, shorter route then became the dominant course.

The timing of the formation of this second channel, and possibly even an additional 'cross channel', is not agreed upon. Some propose it was before the Romans constructed their fort and first settlement, Glevum, here in the first century CE, whilst others argue that it was after this. At some stage, possibly during severe flooding in the fifteenth century, Hafren once again overflowed her banks at the start of her meander, creating a third channel to the west. The original, most easterly channel subsequently dried up, leaving us with the two courses at Gloucester that we know today. The newest, most direct and wider route is the one that passes to the west of Gloucester, and the narrower, now the eastern channel, passes beside Gloucester docks.

As early as the late Iron Age, Hafren connected the Celtic Briton Cornovii tribe, who controlled Shropshire as well as Cheshire and the Welsh hills, and the Dobunni tribe, who controlled the area around modern-day Cirencester and Gloucester including Hafren's final section to the sea. The tribes hollowed out log boats from single tree-trunks, this mode of transport being quicker than overland travel. The Romans founded Glevum here early in their occupation of Britain, attracted by the iron deposits in the nearby Forest of Dean that could then be transported onwards by boat. Its location also provided a defensive position against Welsh incursions at what was known to be the lowest point to cross Hafren. By the second century CE, Glevum had a forum, baths and a basilica as well as stone ramparts to replace the earlier clay ones. The Anglo-Saxons fought over control of this defensive point; in 679 CE King Osric established an abbey here, and in the ninth century Anglo-Saxons began to settle in numbers.⁵⁵ Gloucester continued to develop as an inland port, even getting a mention in the *Mabinogion* story of 'Culhwch and Olwen'. The dubious moral character ascribed to the town in the story perhaps reflects the reputation of ports all over the world.

Early in the quest, Culhwch's kinsman King Arthur sends several members of his court, including Gwrhyr Gwalstawd Ieithoedd, to rescue Mabon, son of Modron, who had been taken from his mother when he was three nights old. Gwrhyr's special skill is that he can speak many languages, including the language of animals. First, they travel to speak with the ancient Blackbird of Cilgwri to see if he knows of Mabon's whereabouts. The Blackbird takes them to speak with a more ancient creature, the Stag of Rhedynfre, who in turn takes them to the wise Owl of Cwm Cawlwyd. The Owl takes them to speak to an even more ancient creature, the wise Eagle of Gwernabwy, who in turn guides them to the wisest, most ancient creature of all, the giant Salmon of Llyn Llyw at a place many believe is near present-day Chepstow. The Salmon tells them that he has travelled up to Caerloyw, (literally, the 'castle/fort glowing bright', and the Welsh name for Gloucester) on the flood-tide and calls it a place of much wickedness. He has learnt that Mabon is imprisoned there, so he tells two of the men, Gwrhyr and Cai, to jump on his back, and they ride up to Gloucester together where they manage to speak with Mabon through his prison wall. The men go to Arthur to tell him where Mabon is imprisoned, and Arthur and his soldiers then manage to free Mabon. This story highlights how salmon are ancient, wise creatures that have been on planet earth much longer than humans. They continue to communicate and share their wisdom with us today. Declines in their number provide important warning signs of increasing pollution levels, and seeing a wild salmon still calls to us as a moment for celebration.

In 1580, Elizabeth I granted Gloucester the status of a customs port. It had its own custom house so that goods could be landed here and import tax calculated and collected. It was the most inland

Hafren

customs port in the UK and had the advantage of reducing the need for slow overland travel. Despite this, the port began to lose trade to the ports of Cardiff and Bristol, due to the treacherous tidal waters between Gloucester and Sharpness. In 1793, the UK Parliament gave permission for the 16-mile- (26-km-) long Gloucester and Sharpness Ship Canal to be built so that shipping could avoid this section. The canal and main basin in Gloucester were opened in 1827, and docks were also constructed at the canal's southern end where it joins Hafren at Sharpness. When the Victoria Basin in Gloucester was added in 1849, the port became even busier, often with thirty tall ships as well as barges and small craft loading and unloading at the same time.[56] It is around these canal wharfs, rather than Hafren, that the imposing bonded warehouses, where goods from overseas could be stored awaiting payment of tax, are clustered. The building of warehouses continued until the 1870s. The Gloucester port remained popular until the 1960s and the docks at Sharpness still operate commercially today.

We visit Gloucester on a cool but, thankfully, drier day than the one we had chosen for visiting Deerhurst a few weeks before. Weak sunshine occasionally breaks through the cloud cover. Arriving in the docks, I am amazed by the sheer size and number of the warehouses from the nineteenth century that surround the wharfs and stand as testimony to Gloucester's long history as an important port. Standing beside the large Victoria Basin, Peter and I look up at these huge, recently renovated red-brick 'cathedrals of trade', now converted into restaurants, bars and apartments. We read aloud some of the names emblazoned in large letters just beneath the pitched, tiled roofs – Victoria Warehouse, Britannia Warehouse, Albert Warehouse. There are many more. Each one has between four and six storeys with small windows evenly spaced along the length of every level. From an information panel we discover that the windows were for ventilation rather than light, and instead of the glass we see today the windows were covered by slatted shutters. The warehouses were mainly built to store 'corn', a generic term covering

Feeling the pull of the tide, the pull of home

wheat, barley, oats, maize, linseed and cotton seed. This was lifted from the barges into the warehouses using winches positioned in the lofts. Now it is pleasure crafts and brightly coloured canal barges that bob slightly in the water rippled by a chill breeze. The reflections of the warehouses shift and shimmer. I find the atmosphere slightly eerie, all these windows looking down on us, the stillness of this once hectic, noisy place now only disturbed by the occasional raised voice or the excited cry of a small child scooting by.

Passing in front of the former Custom House and the port's original Main Basin, we follow the signs for Hafren. We use the small Lock Bridge to cross over the narrow human-made channel that links Hafren to the canal's Main Basin, built to enable smaller boats, filled with corn and other goods, to continue their journey upstream. In front of us is a small, low building, its higgledy-piggledy roof sitting low on old, irregular brick walls that cling to Hafren's banks: a great contrast to its tall, regular, restored neighbours. It is now a shop selling vintage furniture and small antiques. Chatting to the charming, knowledgeable owners we learn that the building is over 200 years old and was originally a well-placed, busy ropeworks.

As we take a footbridge over Hafren, we pause to look back at the old building reaching down towards the green-brown water of her eastern channel. Willows and the now-familiar pink flowers of invasive Himalayan knotweed line the banks. Leaving the bridge, we step onto Alney Island, a flat, low, expanse of alluvial hams (water meadows) prone to regular flooding. This southern end is now a nature reserve, cared for by Gloucester City Council using conservation grazing, mainly by hardy Old Gloucester cattle. Brambles line the path, their berries still green for now, but holding the promise of dark juiciness. Before us, rough grass stretches out to a stand of trees. This pastoral scene appears ancient, but I have learnt that this is not the case. Hafren's course has danced across this landscape, flooding, receding, creating new channels. Humans have built boundaries and barriers around her and across her, but nevertheless she still finds new paths. New land will form, areas may become submerged, islands will stand between new channels. Routes of many kinds that today appear fixed and solid can and will change.

It is a warm summer's day when, a few weeks later, Peter and I visit Framilode, a small village on the edge of the narrow Arlingham peninsula, surrounded on three sides by an elongated meander in Hafren's course. Just a few miles south-west of Gloucester, the village nestles in the broad Severn Vale, an area of fertile land that Hafren has helped to shape on this final stretch of her journey homewards. To my untrained eye, it is from here onwards that Hafren begins to take on what I can recognise as the wide form of an estuary, with mudflats exposed at low tide. As I stand here on Hafren's bank, wispy clouds float above the wide expanse of sparkling, tidal water. I look across to the muddy banks opposite and the rolling green farmland with May Hill in the distance. It seems idyllic now, but at low water the treacherous mud-flats, which so misled the newly arrived Anglo-Saxons pursuing the Celtic Britons over a thousand years ago at nearby Priddy Point, appear. This place, now the epitome of tranquillity, was also once busy with river transport. It was here that trows could join the Stroudwater Navigation canal that opened in 1779 to provide a link between Hafren and the Cotswold town of Stroud, then travel onwards to the Thames at Lechlade via the Thames and Severn Canal completed in 1789. The trow painted on the sign of the village's Ship Inn reminds us of the trading past of this peaceful spot beside the now-redundant first section of the canal, where moorhens swim amongst the thick reeds.

We pass in front of St Peter's, a large Victorian church protected by a high flood-bank. Leaving Peter there, I climb over a stile and drop down onto a riverside footpath. This runs between the raised, defensive bank and a meadow of tall grasses, interspersed with areas of brambles and the tall, purple, thistle-like flowers of common knapweed. I am feeling hot and somewhat weary at the end of a day exploring but my spirits lift as I see so many butterflies fluttering around me. I sit and watch their dance: small whites, their white forewings edged with a black tip and highlighted by a single black spot, and commas, their frilly edged wings opening to reveal vivid orange and brown zigzags. I think a red admiral flits past, but it is too quick to for me to be sure. I enjoy them all, not to classify them but for the pleasure of being with them – the largest number of butterflies I think I have experienced in a single spot. The high grass and sunken path create a perspective reminiscent of my childhood where things stood tall above me. The moment feels magical. Yet there is a danger in this, too. The nature writer Melissa Harrison discusses this using the concept of 'baseline shift'. Each generation becomes accustomed to seeing fewer butterflies, fewer insects, fewer flowers, fewer trees, fewer fish, less wild space, less... In nature writing from the early twentieth century, she has noticed the use of words such as 'clouds', 'shoals', 'flocks', 'swarms', to describe the abundance of wildlife, some of which she has never even seen. I am excited to see a dozen butterflies in one place, but at one time, could I have seen clouds of butterflies here? There is a danger that we become accustomed to a certain lack of abundance or imagine that plant and animal life still exists at the same level as in our childhood memories.

Butterflies and moths are in a state of huge decline. In the UK, 80 per cent of butterfly species have decreased in abundance or distribution or both since the 1970s.[57] Butterflies, such as the northern brown argus and the small pearl-bordered fritillary, which need specific habitats such as flower-rich grassland, heathland, woodland clearings and damp sunny areas, have fared worse than butterflies, such as common whites, that can live in a wider range of countryside and urban settings. The UK's Butterfly Conservation charity (which also helps to care for moths) emphasises, however, that it is not all bad news. For example, the brown-and-yellow chequered skipper, a butterfly conservation priority species, has expanded its range in Scotland. In addition, although previously extinct in England, it has, since 1976, started to breed again in Northamptonshire, thanks to an ambitious conservation project.

The decline in butterflies mirrors species loss across the UK caused by many factors including climate change, intensive farming and land-management practices. Research highlighted in 2023 indicates that in the last fifty years, 38 million birds have vanished from UK skies, 97 per cent of UK wildflower meadows have been lost since the 1930s, and 25 per cent of UK mammals, including water voles and mouse-eared bats, are at risk of extinction. Overall, 40 per cent of UK species populations have declined and the UK is in the bottom 10 per cent globally for protecting nature.[58] These are devastating statistics. Reading them can lead to feelings of despair, powerlessness and a questioning of whether it is even possible to be at home in a world where so many have lost their homes. It can lead to *hiraeth*, a well-known, widely examined, impossible-to-translate Welsh word that invokes a sense of nostalgia for Wales, or for home more generally. It also contains a sense of grief and yearning: a homesickness for what is just out of reach, a longing for a lost world, for the way things were or the way one imagines things to have been.[59]

Hiraeth can trap us in melancholy, into being dismissed as dreamers and, more dangerously, into localism or nationalism where change is seen as a threat. Svetlana Boym's thinking on *restorative* and *reflective* nostalgia is helpful here.[60] Recognising the fluidity between the two 'types', Boym highlights how *restorative* nostalgia often arises in times of crisis. It is characterised by a desire at an individual and/or a society-wide level to reinstate 'a truth' based on a return to origins and a firm idea of how things once were. In *reflective* nostalgia, however, there is no single notion of truth or desire to reinstate a 'true past'. Instead, *reflective* nostalgia 'dwells on the ambivalences of longing and belonging'; it cherishes fragments of memory whilst at the same time it examines them critically, questioningly and with compassion. *Reflective* nostalgia recognises how fantasies of the past are influenced by situations in the present and this, in turn, has a direct impact on the 'realities of the future'. Future actions can be informed but not constrained by notions and methods of the past. The regenerative farming scheme introduced by Alexander Newport at the Bradford Estates in Shropshire springs to my mind here. Alexander and I have had tremendously different lives yet, when we met, we could connect over cherished memories and experiences from our childhoods: the park and farmland, the village school, my meeting Alexander's grandfather walking in the woods he loved. Regenerative farming draws on such memories and the knowledge and skills of the past but puts them to use in new ways. Some methods, such as reducing soil tillage and using a variety of new high-tech machines, will no doubt be questioned. Lessons will be learnt, adjustments will be made, but these are to be expected and welcomed as part of a reflective process.

As Hafren continues along this final stretch of her long journey homewards, the distance between her banks widens further. At low tides, her watery thalweg (central course) is visible, carving a meandering route through the glossy mudflats. These mudflats are bordered by salt marshes, regularly washed by Hafren's brackish water where concentrations of salt vary according to the tide. This creates habitats for salt-tolerant plants such as samphire, sea purslane and purple sea aster, with sea lavender growing on drier areas. These muddy, marshy zones – part land, part water – are central to the work of the Wildfowl and Wetlands Trust at Slimbridge, approximately 4.3 miles (6.9 km) south of Framilode as the crow flies. Founded in 1947 by Sir Peter Scott, the reserve covers 325 hectares (803 acres) and is designated a Site of Special Scientific Interest (SSSI). It provides a home for a wide range of plants and creatures, ranging from insects such as damsel- and dragonflies, to frogs, toads and newts, small mammals and many different birds. Sir Peter Scott described Hafren's banks here as an 'avian Serengeti' providing a home for 30,000 overwintering birds that feed on the rich mud and pasture where a square metre of mud, 2.5 cm (1 in) deep, contains as many calories as thirteen Mars Bars. The hungry birds include Bewick's and whooper swans, Canada and white-fronted geese, redwings, golden plover and pintail and

Hafren

wigeon ducks. Perching birds such as kingfishers, willow tits, reed warblers, buntings and curlew also make their home here, many of them year-round residents.

Peter and I visit Slimbridge on a warm summer's day. A light breeze ruffles Hafren's surface and fluffy clouds scud across a bright-blue sky as we follow the footpath along Hafren's bank. The reserve is set on a wide bend and Hafren and her mudflats combined are about 2 miles (3.2 km) wide. Tall rushes sway gently on the banks, the peaceful scene belying the strength of Hafren's floodwater, which can quickly spread here. We climb the observation tower overlooking scrapes (shallow pools) and the extensive salt marshes and mudflats, and watch the avocets as they step delicately through the shallows. Their black-and-white plumage and long, upturned bills make them easy to identify. There are few birds around as we relax in the gentle atmosphere. As we sit here, I recall the hectic scene we had witnessed on a visit made previously on a cold, bright winter's day.

'Have you seen the cranes?' were the words on everyone's lips in the observation tower that day. In the distance, we could make out the shape of three large cranes, almost ostrich-like in their appearance, amongst the thousands of geese that were moving like an army across the grassland, heads down, intent on feeding, alongside Bewick's and whooper swans and pochard ducks. Through our binoculars the details of the cranes sprang to life with their long legs, trailing tail feathers and upright stance making them clearly distinguishable. It felt amazing and humbling to see these wild creatures. Cranes were once widespread in the UK but became extinct in the sixteenth century due to hunting and loss of their wetland habitat. In 1979 a small colony re-established itself in the Norfolk Broads and then spread to other parts of the UK, aided by the creation and improvement of wetland habitats. In 2010 the Wildfowl and Wetlands Trust started a project to reintroduce cranes to the Somerset Levels. Over a five-year period, over 100 eggs were taken from wild crane colonies in Germany (where cranes are more widespread) then hatched and raised at the Trust's specialist breeding centre at Slimbridge. The staff fed the baby birds using crane-head puppets so that the young cranes would not become imprinted on their human carers and attached to the reserve. Using these puppets, the staff also taught the youngsters to socialise with each other and to forage by pointing out small creatures for them to peck. They even taught the young birds to recognise predators by playing animal sounds and crane alarm calls and then role-playing the crane heads fleeing from danger. The cranes were successfully introduced into the Somerset Levels. A few, however, decided to return permanently to Slimbridge despite the role-playing precautions taken to discourage domestication. Others have chosen to have two homes, travelling between Slimbridge and the Levels, using both the M5 and Hafren's lower reaches for navigation. It is these cranes that we were so lucky to see on the day of our visit. Towards the end of our time in the observation tower, a five-year-old girl entered with her grandfather. We told them about the three cranes near Hafren's bank. She looked, with the naked eye, across to where we were pointing and said: 'There are two more further back.' Sure enough, when we looked through our binoculars, we could see these, too, and the excitement spread around. Five cranes.

To travel home from Slimbridge, we have to go over the Gloucester and Sharpness Canal, and we decide to return another day to explore the canal's southern end at Sharpness. When we come back a few weeks later, looking around the docks is a surprise. Sharpness retains the bustle and slightly shabby look of an international working port, which is absent from Gloucester docks. The volume of trade, however, has significantly reduced in recent years as the accessibility to the port is significantly impacted by Hafren's strong and widely varying tides. The port is believed to be the biggest inland port in the UK. It also operates as a dry dock and centre for repairs. Crossing several metal footbridges, we follow the footpath that weaves its way through the docks and skirts around basins and metal fences. We pass boats of all different shapes and sizes, with that ungainly look boats have when out of the water, and several

Hafren

tall wind turbines. We soon reach the broad canal connecting Sharpness to Gloucester. Here we see the two stone piers that are all that remains of the Severn Railway Bridge destroyed in a fire when two barges carrying oil and petroleum collided with it in fog in October 1960, with the loss of five lives. The bridge connected the port with the coalfields in the Forest of Dean, one of the main goods transported from Sharpness. The loss of the bridge also divided communities living on Hafren's two banks. Children could no longer quickly travel to school using this route, a loss noted by a friend, who recalls this separation, including how this reinforced the pain of the psychological separation he experienced as a boarder on the opposite bank from his family.

We set off northwards along the footpath that runs along the bank of the canal closest to Hafren. Built as a ship canal, approximately 26.4 m (86.5 ft) wide and 5.5m (18 ft) deep, it is completely different from the narrow Montgomery Canal we walked along at Welshpool. I peer over a stone wall covered in brambles. Hafren's wide sandbanks, mudflats and watery central thalweg fill the landscape under a cloudy sky: her wildness and curves contrasting with the arrow-straight canal. Big skies and peace reign here in contrast to the industrialisation at the nearby port. Birds fly low as they head towards Slimbridge, a few miles north. We soon reach an area known as the Purton Ships' Graveyard. Forces of erosion naturally make Hafren want to shift eastwards on this bend, towards the more direct route now occupied by the canal. In the twentieth century, barges, schooners and trows, redundant when trade shifted to other more easily accessible ports and to the railways, were beached unceremoniously here to shore up her left bank. The last boat was beached in 1965. Some of the boats are covered in silt leaving little visible, but here and there we can see a hull or the remains of a wheelhouse. Recently, interest in these boats as preserved monuments has grown. *Harriet*, the only known surviving Kennet barge, beached here in 1964, has now been scheduled as an ancient monument. This designation has lent the other boats and the whole site historic status: a memorial to the significance of Hafren's trading history and the boats and boatmen who worked on her. Black metal plaques with white borders and lettering have been placed beside some of the hulks, giving the impression of gravestones. We stop next to the barely visible remains of *Society* and read that she was built in 1906 in Gloucester, was 24.7 m (81 ft) long and 5.8 m (19 ft) wide and beached here in 1956.

At the Purton canal-boat mooring zone, the banks are suddenly bustling with activity: people washing dishes on deck, mending, chopping wood, chatting to their neighbours and three children playing on their bikes. As we watch, the children throw the bikes down on the path and jump onto a boat to call for a friend. Now four, they climb down from the boat, struggling slightly as they lift an additional bicycle with them. They stop and speak to us. They tell us they live full time on the canal and ask us what we are doing and where we are going. We explain and wish them an enjoyable day. We watch as they set off to pedal energetically along the canal path and over the narrow bridge that leads into the woods on the canal's opposite bank. I know that these children, like all children, have their own particular issues to deal with, but they have a certain joy and lightness of being we can but envy. I wondered what it must be like to live full time in and on this watery world.

As we arrive back at Sharpness, the view of the setting sun over Hafren is amazing. Dark clouds dominate above us but in the west the sky is filled with golds and oranges. With her broad expanse of mudflats, there is no doubting the tidal nature of Hafren here. We can see down to the two bridges that cross her near Chepstow and Bristol. Also outlined on the horizon are the two, now-decommissioned, nuclear power stations at Berkeley and Oldbury. Their imposing twin cooling towers give them a castle-like appearance. Hafren's tidal waters were used for cooling here and also at two stations, now decommissioned, at Hinkley Point on Hafren's broader estuary in Somerset. A new power station, Hinkley Point C, is currently under construction. The ecological impact of nuclear power is much debated, and there are current concerns about the fish that will lose their lives as part of the nuclear cooling processes at Hinkley Point C, as well as general nuclear pollution. Plans to harness Hafren's power using a tidal barrier for the generation of 'green' energy (a disputed concept) have also been widely debated over the years, but so far plans have been rejected because of the impact on this internationally recognised SSSI. It is renowned for the overwintering and passage of wading birds, and seven species of migratory fish move through this area including large numbers of Atlantic salmon as well as shad, twaite shad, sea trout, eels and lampreys. The area's extensive networks of drainage ditches filled with fresh water and brackish water (where seawater and fresh water mix together, and the concentration of salt varies with the tides) would be destroyed by a tidal barrier. This would threaten the survival of mammals such as water voles and also aquatic invertebrates. The mix of sandbanks, mudflats, rocky beds, sea-grass beds and biogenic reefs constructed by worms enables a wide variety of plants to flourish including glasswort, eel-grass, common reeds and sea barley. Nationally important rare species, such as the bulbous foxtail, would also be affected. As we go into the future, there will be increasing pressure to 'sacrifice' this diversity for energy production, with some estimates suggesting that Hafren's tides could generate up to 7 per cent of the UK's energy needs. Others suggest that this is an overestimate and ask, in any case, why human wants should take precedence over the needs of the rest of the planet's inhabitants.

Much of the land bordering Hafren here around Sharpness has long been owned by the Berkeley family. They have taken an interest in the ecology of the area and supported the development of the Slimbridge reserve. Although we cannot see it from this viewpoint, we know from previous visits to the area that Berkeley Castle is tucked slightly inland away from the power station. William FitzOsbern built a motte-and-bailey castle here in 1067, shortly after the Norman conquest, as a Marcher castle to defend

Feeling the pull of the tide, the pull of home

the Normans from both the local population and the Welsh – once again, Hafren playing a role as a border. After various changes in ownership, the castle and estate ownership passed in 1152 to the founders of the Berkeley family who still live there. The town of Berkeley grew up around the castle and it is still a bustling market town today. This is the place where Edward Jenner famously conducted smallpox vaccine experiments. Aware of the protection that local dairymaids acquired when they contracted cowpox, he experimented with vaccinating humans with some cowpox virus. He then exposed them to smallpox, which they did not then develop.

It is getting late and time to leave Sharpness, but we have had a beautiful day here. Several couples are walking their dogs, and one woman gets up from a bench. She tells us she often comes over from the row of cottages opposite to watch the sunset: the broad horizon and ever-changing water here always lifts her spirits. They lift mine, too.

Several weeks later, we decide to return to nearby Oldbury-on-Severn, 7 miles (11 km) downstream from Sharpness, to get a closer look at the power station. Situated in the Lower Severn Vale Levels, it feels very remote. Once forming part of the wider Somerset and Avon Levels, it has become cut off from them by the expansion of Bristol and the industrial developments at Avonmouth. The criss-cross of lanes reminds us that this area has long been inhabited: a strategic point on the mouth of a major river. This sense of longevity is reinforced in the village, fringed on its eastern edge by Oldbury Camp: a double-banked Iron Age marsh fort covering 4 hectares (9.88 acres), known locally as the Toot. The village has also retained the use of the old word *pil* for the tidal inlet running from Hafren up to the village. We take the grassy footpath beside the narrow *pil* towards Hafren and pass a large, metal floodgate that prevents surges of water travelling up the *pil* at high tide. Beyond this gate, the ground is increasingly soft beneath our feet and, arriving at Hafren's steel-coloured water, we can see that the dark brown mud on her banks is several metres thick. The two bridges that cross Hafren downstream dominate the skyline, and the fresh wind blowing upstream rustles the tall grasses and teasels. There is no doubting the pull of the tide and our nearness to the sea. In 1885, a whale made it this far upstream, before becoming beached at nearby Littleton Pill. This caused an immense stir, and visitors came from far and wide to see the unfortunate creature. Looking northwards, we can see the Thornbury Sailing Club building and the tall masts of small yachts. Beyond that looms the silent, forbidding Oldbury Nuclear Power Station, undergoing the lengthy decommissioning processes needed when nuclear power stations become redundant. We stumble a short distance along the Severn Way, but the soft ground and tufty grass is hard going, and we are happy to turn back to the village and the attractions of the pub, the Anchor Inn.

This area of land, extending across to the *pil* at Littleton, is managed as salt marsh, using cows for conservation grazing. The aim is to create vegetation of different heights as well as small areas of bare land. This encourages a diverse range of plants that, in turn, support insect and bird life. Grazing patterns are timed to allow plants to grow and seed, and regular movement of cattle ensures over- or undergrazing is avoided. Cows have long been an important part of agriculture on the Severn Vale Levels, and the Old Gloucester, a dark mahogany or black cow with white flecks along its backbone, was traditionally favoured in this area: good for the production of meat, milk and cheese as well as strong enough to be used for hauling goods. Cheese production was important in this area, especially as the area's remoteness meant that milk sometimes could not be collected from farms. The distinctive golden-red colour of the double Gloucester, a semi-hard cheese, was traditionally created using *Galium verum* (lady's bedstraw) but is now provided by the natural additive, annatto, made from the seeds of the achiote tree native to tropical regions in Central and South America. Double Gloucester is creamier and less crumbly than the other local cheese, the single Gloucester, which has a lower fat content and is paler in colour.

Feeling the pull of the tide, the pull of home

After lunch, which sadly does not include Gloucestershire cheese, we set off to visit St Arilda's Church, perched on high ground half a mile south of the village. This circular, raised hillock is widely believed to be a tumulus (an ancient, possibly Bronze Age, burial chamber) and Roman coins have been found here, too. The location is certainly striking, with the church elevated in this flat flood-plain, and the tumulus has long been a landmark used by boatmen on Hafren. St Arilda is believed to be a Saxon saint, beheaded by the tyrant Muncius for refusing his 'unwelcomed advances'. She was buried next to the spring where she had lived, but her bones were later exhumed and moved to Gloucester Cathedral. Tradition says that the spring water here turns red and stains the surrounding stones with her blood, but a more prosaic explanation is that this is caused by the alga *Hildenbrandia rivularis*. Local people now make an annual pilgrimage from the church to the spring on her saint's day, 20 July. Today, as we enter the church, I am immediately struck by the light. The windows are made from clear glass edged with narrow panes of yellow and pale blue stained glass. These fill the white-painted church with a golden glow as light floods in from the sun setting towards the west. The traditional Victorian mahogany furnishings contrast with more modern features: stacks of chairs, displays of artwork from the local primary school, and headphones to listen to the voices of local people recorded as part of a living history project. The effect is a little untidy, but in a good, lively way. I am immediately drawn to the windows and gaze out. The stone lintels and coloured panes create a beautiful, golden frame for the bird's-eye view that the church's position affords. To the south-west, the flood-plain spreads out, green fields edged with trees and hedges. Hafren's wide water meanders across this flood-plain she has helped to shape over millennia, and beyond this are the hills of Wales. I can feel the deep ancient connections here beneath my feet and carved into the landscape as Hafren feels the pull and push, the rise and fall of the tide as she nears home and her destination, the sea.

Chapter 11:
Reuniting
Minsterworth, Gloucestershire, to Aust, South Gloucestershire, and Chepstow, Monmouthshire

As the days shorten and blackberries ripen, interest in tide times grows in the Severn Vale. The high tides around the autumn and spring equinoxes are the best times of the year to experience Hafren's 'final surprise': the Severn Bore. Hafren has one of the world's highest tidal ranges, with other high tidal ranges including those recorded at the Bay of Fundy (North America) and at Ungava Bay (Hudson Straits, North America). This tidal range – the height difference between low tide and high tide – can be as much as 15 m (49 ft). Combined with the funnelled shape of Hafren's estuary, this can lead to the creation of a visible wave of water, travelling up-river as far as Gloucester at speeds of between 8 and 12 miles (13 and 19 km) per hour. As the river narrows, the height of the wave increases. Tides of over 9 m (30 ft) at Sharpness will normally produce a good-sized bore, although tidal ranges over 10 m (33 ft) produce what is often called a 'five-star bore'. Changes in barometric pressure and high rainfall (creating an increased flow of downstream water) can reduce the size of the bore, and strong winds from the south-west can increase it. My friend Kate is keen to see the bore, so together we consult the tide timetables that the internet makes so accessible these days. The highest bores are predicted to be during moonlight hours, but there is a late-September date with a moderately high daytime bore and a good weather forecast.

It is first light when I make my way down to Kate's house. Equipped with flasks we set off by road, following Hafren's route northwards. I think of the people who have headed north in earlier times using log boats, trows and barges and also the cranes journeying between their two homes – the Somerset Levels and Slimbridge. As we pass the turning for Slimbridge, Kate tells me something about bird migration she has recently learnt. Throughout history, people wondered what happened when some species of birds 'disappeared' during wintertime, and came up with various ideas. In his *Historia Animalium*, written in the fourth century BCE, Aristotle did correctly identify that some birds migrated. However, he hypothesised that others hibernated, suggesting that swallows, for example, spent the winter in rock crevices. He also proposed that some 'winter birds' and 'summer birds' were the same birds in different plumage. Since then, various thinkers have explored this conundrum, proposing that birds dug holes and buried themselves until spring or that they shapeshifted into other creatures, such as mice or voles, for the winter, undergoing metamorphoses in the same way that caterpillars change into butterflies. The idea that birds could fly vast distances to spend winter and summer in different parts of the world was widely considered preposterous and discounted. It is only relatively recently that bird migration has been better understood. In 1822, a stork caused a stir in Mecklenburg, Germany, when it arrived pierced through its neck by a 76-cm (30-in) wooden spear with an iron tip. The stork was taken to the University of Rostock and became known as the Rostock *pfeilstorch*. At the university, the botanist Heinrich Gustave Flörke analysed the spear and deduced that it was made of a tropical wood used for spears in the upper Nile region. He concluded that the bird must have overwintered there before migrating to Europe for the summer.

Although the Rostock *pfeilstorch* is widely considered to be the first tangible 'proof' of bird migration (and further storks pierced by spears were subsequently recorded), it did not immediately change views on bird migration. Instead, understanding grew over time as increasing travel allowed for better observation of migrating birds, and the practice of placing coded aluminium bands on birds' legs increased the data available.[61] Recent developments, such as satellite tracking, analysing birds' diet and even their changing gut biomes, provide increasing amounts of granular information on bird migration patterns and how, for example, these are affected by climate change. There is also now potential to try to understand more using AI. I think of the tiny willow warbler that I heard at Porth Farm, weighing only 9 g (0.3 oz), flying a huge distance to its winter home south of the equator. It does seem incredible. As Jamie, the scientist at the fish pass in Worcester, had discussed with me, there is still so much about bird, fish and animal migration we don't understand and have still to learn.

Kate and I skirt around the westerly edge of Gloucester, and I can see the roofs and upper storeys of the tall warehouses at the Canal Quay. We cross Hafren at the aptly named Over Bridge, then head a few miles back south to Minsterworth, on Hafren's right bank. This is a favoured spot to watch the bore as access to Hafren's banks is easy and, as she narrows and curves, the effect of the bore is intensified. Our journey has been quiet, as one would expect given the early hour on a Saturday morning, but suddenly we are in a lane busy with people. There is no doubt we have come to the right place. We follow our fellow early risers, pass in front of the small village church dedicated to St Peter and walk up the steeply raised banks, built to protect the village from flooding. People of all ages already line the top of the bank, many of them settled in folding chairs, eating hearty breakfast picnics. We find a spot and, looking across Hafren, we can see a grassy meadow and a farm, its barn filled with hay ready for the winter. In the distance, the purple-blue outlines of higher ground are set against a sky of soft yellows and greys. Hafren's brown water, rich here with silt, reflects the trees that punctuate the meadow. A surfer and a paddleboarder cross the field, clamber down the steep, muddy bank into the water and position themselves in anticipation of catching the bore. As we wait, we listen to the sounds around us: the chatter of children asking, 'Is it coming soon?', the hum of humans chatting and birdsong. I can feel the damp from the grass seeping through my shoes and I am happy I can wrap my hands around a warm mug of coffee. We wonder if we will definitely spot the bore or if it will be an event that happens and one is not sure if one has actually seen it.

Suddenly, here it is. There is a collective excited 'Ahh, look' from the crowd. 'Well, there's no missing that,' I exclaim. Even though I had read about this phenomenon, it still takes me by surprise. As the bore approaches the bend, we can see the frothy, white-crested wave, several feet high, leading the way and acting as a wall in front of a large body of swirling water. What Kate and I hadn't expected was the roar, the sound of the sea brought inland. As the crested wave rounds the bend and passes in front of us, Hafren becomes a turbulent, undulating mass. Silky-smooth, criss-crossing waves ripple, glisten like rich melted chocolate and lap high up her banks. The energy in the water is palpable. We hadn't expected this. In all that we had read about the bore, the emphasis had been on the wave at its front, rather than this powerful water that travels up-river with it. As I stand here, I think of the story of the giant Salmon of Llyn Llyw bringing Gwrhyr and Cai to Gloucester on his back. Some commentators suggest that this is a reference to the men riding the bore up to Gloucester. The water in front of me really does remind me of a lively, leaping, twisting salmon with glossy scales.

The leading section of the bore passes, travelling onwards to Over Bridge at Gloucester, but the water level and the energy caught up in it remain high. Kate, a paddleboarder herself, tells me she was watching the two men in the water as the bore travelled upstream. The surfer, having missed the crest of the wave, returned disappointed to the opposite bank and climbed out. The paddleboarder, however, had managed to climb onto his board before the bore arrived and paddle on the higher turbulent water it had brought in from the sea. His task was made even more challenging by the need to avoid the heavy branches swept upstream in the strong currents. I wonder what it must have been like to be in the midst of all that power and motion. Spectators gather their possessions and murmur their goodbyes, but Kate and I are reluctant to leave. We want to spend more time beside Hafren, experiencing the effects of the bore. We walk downstream, following the footpath along the top of a high, raised bank protecting Minsterworth's waterside homes and their long, sloping gardens. We feel confused, however, about the route we are taking now that the direction of the current has changed. It feels as though south has become north.

Admiring a lovely apple orchard, the boughs of the trees laden with fruits including russets – my favourite – we search for windfalls and apples on branches overhanging the path. 'Foraging,' Kate calls it, 'not scrumping.' Hafren is flowing high and trailing willow branches are deeply submerged. I clamber between the birch trees and crouch down beside the alders whose feet are now immersed in the powerful water. Large branches float by, carried upstream by the energy of the bore. Our progress is slow as we both like identifying and enjoying the sights and sounds around us – not to mention the sharp taste of those 'foraged' apples. The songs and calls of chiffchaffs, robins, wrens and crows blend with the nearby traffic noise from the road that also follows the valley to link Gloucester to Chepstow. Rushes stand in inlets in the banks and, in addition to the ubiquitous poplars, willows and alders, we stop to admire the boughs of a mature oak stretching over the path. The autumn colours are late arriving this year, and the delicate silver-tinged leaves of white poplars contrast with the glossy green maples. Kate teaches me to identify ash from the tiny black leaf buds that appear in pairs on either side of the main stalk in the autumn, holding the promise of light, fresh, green leaves next spring. I stop to look more closely at a willow I do not recognise. It has long, dark, narrow glossy leaves on well-spaced, long upright branches, giving it an airy appearance. Using my plant ID app I discover that it is an osier willow, also called basket willow. Its long, straight branches and its ability to quickly produce new growth when coppiced make it ideal for basket weaving – a traditional practice here in Gloucestershire.

With a last, lingering look at Hafren, still running fast more than an hour after the front wave of the bore passed by, we set off home, happy to have shared this special morning and experienced this striking and unusual natural phenomenon together. Our visit was in the daytime, but it sets me thinking of a night-time experience of the bore described by Julie, whom Peter and I had met the previous spring at the ancient, ruined Temple of Nodens at Lydney Park. This is on Hafren's right bank, 15 miles (24.1 km) downstream from Minsterworth. We started chatting with Julie when we all stopped to admire a panoramic view of Hafren through a gap in the tall trees. She told us that she had always lived near Hafren or her wider estuary. Born at Quedgeley, just south of Gloucester on Hafren's left bank, Julie had moved to Portishead on Hafren's wider estuary and was now living on Hafren's right bank in a village near Chepstow. Hafren had always been part of her life and meant so much to her; she could not imagine living anywhere else. She told us how, growing up in Quedgeley, she and her family were a little dismissive of the bore and the tourists who came to see it. That changed, however, one clear night when her mother, Helen, was cycling home along Hafren's deserted banks. The approaching bore, lit up by the moonlight, was so beautiful, so awe-inspiring – graceful yet powerful – that Helen felt she had to stop and absorb the moment. The next day, Helen recounted this experience, and Julie recalls, so many years later, the strong emotion in her mother's voice and her own regret that, asleep in bed, she missed sharing this special encounter.

The ruined temple to the god Nodens, where we have this moving conversation, is set in privately owned parkland only open for a few weeks in May each year. It is just south of the village of Lydney and the once-busy Lydney Harbour, well known for shipbuilding, metalwork and coal trading. The harbour is now over a mile from

Reuniting

the village due to silt forming on Hafren's right bank that has led, over the centuries, to her course moving eastward. To get to Lydney Park we cross Hafren near Chepstow using the 1960s suspension bridge. On the day we visit, the tide is out. Peering down, I can see the glistening mud and the steely water rippling in the crosscurrents where Hafren reunites with her sister Afon Gwy (River Wye) before they fulfil their destiny to rejoin the sea together. We turn northwards at Chepstow, then pass small villages of terraced houses surrounded by fields of verdant spring grass. I can sense that the area has long been inhabited, Hafren drawing people to her over many centuries.

Tall trees outlined on the hill rising from the valley floor announce the entrance to the park. This steep hill provides a long-recognised defensive advantage beside Hafren. Archaeological evidence has been found of an extensive Iron Age promontory fort here covering 1.8 hectares (4.5 acres). In the third century CE, the Romans dug for iron ore, and evidence of tunnels, as well as opencast mines known locally as scowles, can still be found. In the fourth century CE, the Romans built a Romano-Celtic temple dedicated to the Celtic god Nodens, also known in a later period as Nudd in Irish and Llud in Welsh mythology. The name Llud lives on in the current village name: Lydney. Nodens was a god of healing, and several models of dogs, associated with healing in this period, have been found in the area, possibly brought by pilgrims to the site. Setting off to walk up the steep escarpment, we are soon drawn into the special atmosphere of the beautiful surroundings. Over the cacophony of birdsong, the voice of the blackbird rings clear and true. The long, silver leaves of hound's tongue, so very soft to the touch, line the path. Tall trees – lindens, chestnuts, beeches – tower over us, each dressed in their fresh, spring leaf-growth. Next to these huge trees, many 30 m (100 ft) tall, I feel tiny, aware of my insignificance in the overall life of the earth.

Emerging into the clearing at the hilltop, we see the ruins of the temple, excavated by Tessa and Mortimer Wheeler in 1928–9. We walk inside its unusual rectangular shape, now only outlined by the low, ancient remains guarded by a sessile oak standing proud by the entrance. At right angles to the temple, the remains of guest quarters for pilgrims and a bath house are clear, the distinctive pillars that would have supported the raised floor clearly visible. However, on this spring day, what draws me into this place and the woods surrounding it are the tiny colourful flowers – more than I have ever seen before in one place. They nestle amongst the grass, daisies and dandelions at our feet. I crouch down to look more closely at this jewelled carpet – pink miniature geraniums; delicate, blue speedwells and star-like, yellow lesser celandines. The paths and clearings here are seldom trod by human feet. Whilst this does raise questions of access rights over privately owned land, it also makes me reflect on the recuperation of the land that the absence of humans has created here in this temple dedicated to a god of healing.

These recollections of spring are a comfort as autumn turns to winter, and the chill wind tears the few remaining leaves from the trees around my home. But winter has its own beauty, and Peter and I are still drawn to Hafren. Today we are at Aust Cliff, a place we have visited many times. Tucked between the two bridges that cross Hafren between Bristol and Chepstow, Aust Cliff has much to share with those who venture there. Having first checked the tide times to ensure a visit is safe, we follow the Old Passage Road. The narrow lane dips sharply and a wide view of Hafren and her mudflats opens up. We are close to the 1960s suspension bridge and at the place where the ferry ran before the bridge was opened. Peter recalls coming here on a trip with his father in the 1950s, crossing Hafren for a tour through Wales ending at Bangor and Anglesey – quite a journey from their home in south Devon. He cannot picture the ferry, but a visual record is provided by an unexpected source. Bob Dylan took the ferry in 1966 en route to Cardiff after a challenging performance in Bristol, and a black-and-white photo captures him standing moodily on the ferry pier with the almost-complete suspension bridge in the background. Martin Scorsese used this

Reuniting

picture, now widely available on the internet, on the cover of his film *No Direction Home*, released in 2005. We park next to the large cream-rendered Old Passage House, go through a narrow gate and follow a small path down towards Hafren. The tangled brambles beside us are bare but our way is brightened by cushions of ivy and bright bursts of yellow winter-flowering jasmine. We soon join a wide concrete track, built to connect electricity pylons here to road access further south along the banks. A wide bed of tall reeds grows between the track and Hafren, and their brown feathery heads and stems dance in the strong breeze, making me think, somewhat wistfully, of a rippling field of summer-ripened wheat. The concrete track continues to a short pier barred by a gate protecting the pylon and cables stretching out across Hafren to her far bank.

Beyond the concrete, the way becomes extremely muddy. We have to tread carefully and are relieved when we can step onto the rocky 'beach' area and look up at the cliff. Its unusual horizontal stripes tell an amazing story about the long history of the planet. The upper stripe consists of a mix of limestone and mudstone that formed in shallow tropical seas about 210 million years ago during the Rhaetian time interval, which occurred towards the end of the Triassic geological period (252 to 201 million years ago). It is in this layer that fossils of tropical creatures such as ichthyosaurs and plesiosaurs can be found. This 'bone bed' at the top of the cliff is considered as one of the best sites for marine reptiles in Britain. Fossilised bones from these creatures, as well as those of fish such as sharks and the remains of shells and sea urchins, are washed onto the beach below. The wide, reddish-brown central stripe is made of mudstone that formed during the Triassic period when the land that is now at Aust was located near the equator and had a hot desert climate. At the base of the cliff we can see ridges of white. These are soft gypsum rock (calcium sulphate) that formed when desert lakes dried out.

On my first visit to Aust, I was confused when I read on a noticeboard about the cliff's desert origins. How could this cliff, just outside Bristol, have once been part of a tropical desert near the equator? I asked my more scientifically minded sister and her husband, who is also called Peter. He asked me if I had ever heard of continental drift. This did jog my memory. I had heard about the Galapagos Islands moving over millennia, but I had never thought about such things happening nearer home. I read a little wider and learnt about the theory of continental drift and the more recent theory of plate tectonics. The theory of continental drift is most associated with Alfred Wegener, a trained astronomer who drew on botany, geological records of different types of rocks, and fossil records. In the early twentieth century, Wegener proposed that all the continents were once part of a single land mass near the equator, that he called Pangea. He argued that this land mass was forced apart and the pieces began to drift towards the positions they have today. He used geological and fossil records to support his theory. For example, fossils of the mesosaurus, a freshwater reptile 1 m (3 ft) long, are only found in southern Africa and South America. This creature could not have swum across the Atlantic Ocean, suggesting that they once lived in a single habitat with many lakes and rivers. Wegener argued that it was the earth's rotation and the pull of magnetic forces that caused this drift. Some scientists did not agree with this explanation for the movement, and in the 1960s the alternative theory of tectonic plates developed and became widely recognised. This theory proposes that the continents rest on huge slabs known as tectonic plates and these are still moving today. Some of the most dynamic sites of tectonic activity are sea-floor spreading zones under the ocean. In these zones, the movement is caused by lava from the earth's core rising to the earth's crust at the edges of the plates, where it then hardens. This pushes the plates apart and creates new crust. For example, the Mid-Atlantic Ridge is pushing the Eurasian and the North Americas plates apart at a rate of 2.4 cm (1 in) a year. Rift valleys form when tectonic activity occurs on land masses. Today, scientists propose that, over the earth's lifespan, several supercontinents have formed and broken up.[62] I can understand now how Aust Cliff could once have been a desert near the equator, but it still seems mysterious and troubling, challenging as it does commonplace preconceived ideas of the stability of the land around us.

Peter and I step onto the beach, burying our chins further into our coats against the chill. We walk along, eyes down, looking for fossils, our hands getting increasingly muddy as we lift likely-looking stones. We soon find several flat sand-coloured rocks embedded with fossilised shells and traces of corals as well as a large grey stone with a black, rounded addition at its centre. I will take it to show my brother-in-law to see if he can identify what it might be. I miss James, who was with us the last time we came here. He really enjoyed searching for fossils, amazed by the ease with which we could find them. He loved the vast expanse of water, the wide horizon and the closeness to the sea, the feel of fresh breezes blowing in from the west. Raised by the coast, he is always happy to return to such places. Today, we meet Christian and his daughter and son, aged about ten and five years old, all warmly wrapped in bright anoraks. They, too, are searching for fossils, the ten-year-old, particularly, enjoying splitting the stones with her new fossil hammer. I ask Christian if they come here often, and he explains that he used to come more before he had children, but now, they are old enough to come along, too. On one occasion, he even found a fossilised footprint that clearly showed the impression of the creature's claw-like foot. He had taken the fossil to the Bristol Museum where it was analysed and found to be the imprint from some kind of amphibious creature; similar fossilised prints had been found in Nova Scotia on Canada's eastern seaboard.

Sitting on some rocks close to the 1966 bridge, lorries and cars, toy-like, pass above us. I look across to where Hafren and Gwy meet. Here, Hafren's right bank curves towards the midstream, creating a long spit of land. Beyond this, Gwy's waters widen before she merges into Hafren: a reunion of the sisters in a wide watery zone of mudflats, marsh and tides with Wales on one bank and

Hafren

England on the other. I recall the day earlier in the year when Peter and I had visited these merging waters. Buffeted by rain and wind, our feet sliding and sinking into the deep mud, it had been hard to remain upright, immersed all around in a wet, shifting, silver-grey world of mingling waters, histories and journeys. I look more closely at the bridge and how it navigates this challenging terrain. The crossing is complex, consisting of four linked structures. Firstly, there is a viaduct on the Aust side, then a wide suspension bridge across Hafren, another viaduct across the headland and finally a suspension bridge to cross Gwy and reach Chepstow in Wales. I turn and look downstream towards the newer bridge opened in 1996: it is this bridge that now carries the M4, the main road route into south Wales. Silhouetted against the sky, the curving sweep of the whole structure has an elegant beauty. Its 950-m- (3,000-ft-) long central section is a cable-stayed bridge in which the weight of the deck is supported by a number of nearly straight diagonal cables held in tension to one or more vertical towers. The effect is like two pyramids of cables arising from the bridge, pointing upwards towards the sky.[63] The towers transfer the cable forces to the foundations tunnelled deep into Hafren's bedrock through vertical compression. This design makes the bridge less vulnerable to the high winds that blow through the estuary and often cause the older bridge to close. In total, the whole structure, including both approach viaducts, is more than 3 miles (5 km) long.

Although it cannot be seen, I am also very familiar with another important link joining England and south Wales here: the Severn Railway Tunnel. I often travel through this on my way to visit family in Wales and have experienced the loud banging of the carriage windows as the change in pressure caused by our descent sucks them shut. When the tunnel was opened to regular goods trains in 1886, it was the longest underwater tunnel in the world, a record it held until 1987. The tunnel took almost fourteen years to complete and remains an incredible feat of engineering. Its construction by the Great Western Railway, under the auspices of its chief engineer Sir

Reuniting

John Hawkshaw, is even more remarkable when one considers that no laser-guided and tunnel-boring equipment existed at that time. The tunnel itself is 4.35 miles (7 km) long, although only 2.25 miles (3.6 km) of it are under Hafren. As well as the problems of keeping the tunnel as watertight as possible from the brackish tidal water pressing down from above, a significant challenge was the many small springs that the tunnellers struck during construction, that had to be sealed and pumped out. The tunnel was almost completed when tunnellers hit the most significant spring of all, that rapidly flooded the tunnel. Construction was halted for over a year and required the bravery of local diver Alexander Lambert to rescue the project. Using the diving suit and portable oxygen tank equipment newly invented by Henri Fleuss, Lambert was able to close a watertight door and enable this water to then be pumped out. I was surprised and a little unnerved to learn recently that 50 million litres (11 million gallons) of fresh (spring) water – the equivalent of 20 Olympic-sized swimming pools – is still pumped out daily from the tunnel. Much of this is returned to Hafren, but some was used until recently in paper production and some water still makes its way to a local brewery. As you pass through the village of Sudbrook, where the tunnel emerges on the Welsh side, you can still see the cottages and other facilities built for the tunnel workers, and the village's social club has a fascinating small museum. As well as this social and engineering history, it is fun to learn that when the Oasis singer Noel Gallagher was stuck in the tunnel in the 1990s, on his way to the Loco Studios for the *Definitely Maybe* production sessions, he wrote the hit song 'Acquiesce'. The chorus highlights how we both need to and can believe in each other, and through these connections, we will uncover what sleeps in our souls.

Beyond the tunnel and two bridges, the combined flow of Hafren and Gwy widens even more. I had always thought it was here that the two sisters fulfil their destiny to return to the sea, and once again, begin the process of falling as rain as part of the long water cycle. I recently learnt, however, that Hafren's estuary extends further, according

Hafren

to the definition provided by the International Hydrographic Organization, bordered on the English side by the Somerset Levels and on the Welsh side by the Gwent Levels, a low-lying watery zone that hugs the coast from Chepstow down to Cardiff. In this definition, she enters the Môr Hafren (Bristol Channel) at a 9-mile- (14-km-)wide watery boundary drawn between Lavernock Point, south-west of Cardiff, and Sand Point, north of Weston-super-Mare. As well as producing the Severn Bore, the high tidal range and the estuary's funnel shape produce strong tidal streams and high levels of churn in the water. Combined with the underlying rock, gravel and sand, these turn the water greyish-brown. This somewhat drab-looking water has now been seen by millions of young people across the world when it made a surprising appearance in the video for the song 'You and I' that the pop band One Direction filmed at Clevedon Pier near Weston-super-Mare. The tidal range, extending over this wider part of the estuary and inland up to Gloucester, also results in Hafren having one of the most extensive intertidal wildlife habitats in the UK for plants and animals adapted to life in brackish water, where seawater and fresh water mix. Myriad life forms rely on this vast watery wilderness of tidal mudflats, creeks, salt marshes, peatbogs, reed swamps and wet woodland. The whole area is recognised as a Special Protection Area. In addition, some zones, such as those around Slimbridge in Gloucestershire, Brean Down in Somerset and the Gwent Levels upstream from Cardiff, are also recognised as SSSIs. A recent decision against building a relief road for the M4 across the Gwent Levels between Newport and Cardiff was influenced not just by cost but also the damage it would do to the protected, ecologically sensitive, low-lying wetlands and intertidal mudflats near Newport.

The Severn Estuary Partnership has an even wider definition of Hafren's estuary, with the boundary running from Nash Point in the Vale of Glamorgan in Wales to Hurlstone Point, west of Minehead in England. The Partnership, set up in 1995, is an independent initiative that acts as a focus for the activities of various bodies

Reuniting

connected to the estuary, including local government, statutory authorities, farmers, fisherfolk, environmental groups and local people. In 2001, the Partnership published the 'Strategy for the Severn Estuary', which sets out an integrated plan on how to care for this special area. This wider boundary enables more habitats to be included, such as the Wildfowl and Wetland Trust's project at Steart Marshes in Somerset. Much of Hafren's estuary is lined by defensive sea walls but, at Steart, a deliberate breach has been made in the wall on the estuary of the River Parrett as it enters Hafren's estuary. At particularly high tides, this allows seawater to flow into the area next to the breach, creating a salt marsh. This acts as a buffer zone to protect areas further inland from tidal and sea storm surges, something that is increasingly important as sea levels rise. The marsh also provides a home for a wide range of plant and animal life and increases carbon storage in the waterlogged ground. Salt marshes are quicker to establish than forests and can make a significant contribution to carbon sequestration. Peter and I had visited the Steart Marshes earlier in the year after a period of heavy rain and as we squelched our way through the boggy ground we were reminded of the sounds and sensations we had experienced when we had visited Glaslyn lake and nature reserve at Pumlumon. Hafren's widely varying tidal range means that the marshes are not constantly flooded, so farm animals can still be grazed here following conservation grazing practices. It also creates a very particular growing environment for plants and wildlife specially adapted to these unusual conditions. Perhaps the precise boundary of Hafren's estuary is not as significant as an appreciation of this special, watery zone where salt and fresh water merge, where rain falls, where the tides push and pull and where Hafren still rises through the early-morning mists and reaches out to us.

'We have brought you a piece of our crystal,' a high, clear voice sings out beside me. I turn and see that Christian's son and daughter, who we met earlier on the beach, are standing next to me. The young girl is holding out a small piece of gypsum rock and, taking it carefully from her, I thank them both. We bend our three heads over the white stone to examine the fine brown veins that run through it; the outside smoothed by the tides and the rough face where they have just divided the rock with their chisel. We feel the sharp edge this has created under our fingertips. 'The stone is very beautiful, I love it,' I say. The young girl tells me she is Riina, and her brother is called Armand. These are Finnish names meaning 'joy' and 'beloved'. I thank them again for their gift and fold my palm around it. I hold it tight as I watch them skip back towards their father who is standing on the shoreline, Hafren's steely water swirling behind him. The children slow down to clamber over some larger rocks and Riina extends a caring hand to steady her little brother. As I watch them, I try to imagine what a world orientated towards love, joy and care would be like – a difficult undertaking in this world that is so often dark, filled with so much pain, hatred and harm. As we travel into the future, more change will come, both good and harmful. Yet, to recall Rupert Read and his concept of thrutopia, what will count is how we engage together 'to get from here to there' and 'live and love and vision and carve out a future, through pressed times'.

Reflecting

As I sit here today at Aust, with Peter beside me, I reflect on the wonderful, meandering journey I have made with support from family and friends. As I look at the shimmering water and the bridge James and I crossed on our way to Hafren's source, it is hard to imagine that this wide expanse was once a tiny stream emerging from a bog on a remote mountain plateau. My adventures with Hafren have not been undertaken in one go or over a single, unified time frame. Instead, they have been carried out in 'bits and pieces', when health and other circumstances have permitted, but this is ok. Journeys do not have to be a heroic striding over the landscape.

At times during these travels, I have felt frustrated at my lack of independence, my need to rely on others for support. Then, one day, I realised that interdependence is something Hafren is teaching me. In Westernised societies, autonomy is highly valued, obscuring the interdependent web of life in which we are entangled, enmeshed. We cannot live without care and support *for* and *from* family, friends and communities, both human and other-than-human. Care for and from all on this planet has been a recurring theme at every stage of my journey. The artist Otobong Nkanga proposes that such caring is 'a form of resistance'. She highlights how attentiveness to all 'types of life that do not have a voice' can be a starting point to resist 'what the economy has to say, what capital has to say, what politicians decide' about all the inter-connected parts of the earth needed to ensure 'the possibility of existence' in the crucial years to come.[64]

Hafren

My thoughts turn to other things I have learnt on this journey with Hafren. It has highlighted that there is much wisdom in the world we do not yet understand. How do the cranes at Slimbridge know where they were born, even when humans tried so hard to conceal this from them? How do those tiny willow warblers at Porth Farm manage to fly such huge distances and know their way? What draws the pioneer thwaite shad back to their ancient breeding grounds now that their routes, blocked for over 150 years, have been unlocked? Scientists, such as Jamie researching twaite shad in Worcester, are investigating such questions. Scientific research *is* very important but, if we are open to it, there are also other wisdoms that we can respond to. This is the wisdom of ancient salmon jumping through clear water as we delight in seeing them, the wisdom of rivers, intensified at waterfalls, which pulls us to them – if we are allowed access – to pause, reflect, feel, learn.

Journeying with Hafren has opened up new ways for me to appreciate life experiences. Exploring Hafren's watery wide web of tributaries has given me a new understanding and awareness of the tributaries that flow into us, too, building who we are and how we live. I have learnt the value of meandering, both physical, following a slower, indirect route, and also mentally, letting our minds wander, explore, float free – not always an easy thing today in modern, fast-paced societies seeking the quickest solution. Shocks can happen, too, and, just like Hafren, we have to set off on new courses, find new directions and ways to understand and be in the world.

Reflecting, too, on the wisdom of stories, poetry and myths bound up with rivers and their landscapes can help us consider alternatives to the drab, reductionist framing of the world that many, but not all, humans hold, often without realising it. Darwin teaches us that this is possible. Dominant world views can be challenged, and we can be open to different possibilities, even though this is not an easy undertaking. Learning about Robert Owen and his struggles to establish socialist principles and ways of living, and Wilfred Owen's struggles to find his poetic voice and the immense hard work this

Reflecting

involved, even as he faced death, highlights how achieving the outcomes we want in order to live differently in the world will take 'hard graft'. Change to the planet has already come, is happening now and forever will come. The cliffs behind me – once part of a desert on the equator – and Hafren, with her shifting course, her changing flood patterns, teach us this. But how will we respond? How will we move forwards?

My thoughts return to the beautiful journey I have made with Hafren. I reflect on all I have experienced with her as she undertakes the meandering, lengthy route she wishes for in the giant's myth: the water, the air, the soil, the insects, birds and animals (including humans), the clouds, the rain, the squelching mud, the sounds, the stories, the music and poetry... and so much more. I look again at the crystal given to me by Riina and her brother Armand and reflect on their care for each other and their beautiful names meaning 'joy' and 'beloved'. I recall the moments of joy I have experienced on this journey as well as times of sadness. I contemplate the love and care I have witnessed and received, as many people work hard together to help build a future in these difficult times. I feel the love, joy and care that Hafren and the myriad life forms in and around her extend to me and to us all.

Diolch, Hafren
Thank you, Hafren

Acknowledgements

I would first of all like to thank the whole editorial and production team at Calon. I would especially like to thank the Calon publishers: Amy Feldman, for her encouragement at the proposal and early writing stage; Abbie Headon, as my ideas and writing developed; and Caleb Woodbridge, Caroline Goldsmith and Katherine Venn, during the final edits. Thank you to Steven Goundrey, Agnes Graves, Anna Baildon and all in the production, typesetting and proofreading teams. A special thank you to Clare Grist Taylor, my structural editor, and Rebecca Collins, my copy editor: your help, encouragement, ideas and enthusiasm, always so generously given, have been invaluable. I have learnt so much from you both. Thank you, Jon Gower, for the thoughtful and generous foreword. I really appreciate you writing this for me. Thank you, Andy Ward, for the beautiful cover illustration and Agnes Graves for the carefully crafted and informative map of Hafren's route.

I also want to thank all the wonderful people, creatures, flows of water and other elements of the planet that I have met and experienced along my journey with Hafren, and of course here I thank Hafren herself. I have had so many lovely encounters and heard such interesting stories, each one so freely given. I could not have written this book without you all.

Likewise, I could not have written this book without the help, support and encouragement of all my family and friends.

I would like to thank my sister Rachel for the beautiful illustrations – I love them. Thanks are also due to Rachel and to her husband Peter for all their help, especially with the more scientific aspects. However, anything scientifically inaccurate remains completely down to me.

Acknowledgements

I would like to thank my cousins Angela, Eira, Deborah and Alex for sharing their stories about their mothers with me.

Thank you to Mark Shergold for bringing the musical aspects of the work of John Ceiriog Hughes to life for me. I would especially like to thank him for his wonderful piano rendition of Ceiriog's adaptation of 'Codiad yr Hedydd' ('The Rising of the Lark'), which I recorded and took with me to play beside Hafren at Newtown.

Thank you to Kate Nokes for taking me to Minsterworth and for sharing the experience of the bore. Thank you also to Kate and her family for their help and encouragement during the research and writing process.

Thank you to Julia Maclagan and Mike Hames for their help with the Worcester flood records.

Thank you to Alison Harper, Shehana Gomez, Matthew Isherwood, Owain Jones, Kerry Chappell, Jenny Perry, Carol and Patrick Taylor, Chris Turner, Sam Walton, Andrew Spink, Helen Clarke and Sharon Witt for all their help and support with ideas, sharing resources, reading chapters and generally listening to my ramblings and helping me make some sense of them.

I would also like to thank Terry Marsh for his excellent, compact pocket guide *Walking the Severn Way* (Cicerone Press, 2023). Full of insightful gems, tips on the path and useful small maps, my copy is now beautifully thumbed as well as crinkled by the rain it has received during our many visits to Hafren.

Throughout this journey I have used various free apps such as iNaturalist and the Merlin bird app to help me identify species. These are excellent tools available to us all and I am very grateful for them.

Last but not least, I want to thank Peter and James – I could not do the things I do without you both. Thank you for coming on this journey with me, for cheering me on, for endless writing advice and proofreading, for your steadfast presence. You are my constants.

Endnotes

Opening
1. *Encyclopaedia Britannica*.

Chapter 1: Seeking the source
2. Anaerobic digestion contrasts with processes of aerobic digestion that occur in drier areas (with more oxygen available). In aerobic digestion fast-acting microorganisms, which require oxygen to perform their task, cause much more rapid decomposition.
3. *Afon* is the Welsh for 'river'.
4. On their website, the project provides an example of how, early on in their work, they contacted a landowner who had previously consulted them regarding a specific species on her land. They met with her and her farm manager to suggest possible restoration strategies. They also consulted with government environmental agencies since part of the area identified for restoration was a Site of Special Scientific Interest (SSSI). Initial misgivings were soon replaced as the changes made, for example the blocking of drains to re-wet the peat, quickly improved biodiversity and also provided economic benefits. Other landowners and farmers in the area heard of the successes and became involved, creating on-the-ground demonstrations of success. In pilot areas, farmers are paid on average £265 per hectare (£107 per acre) to restore peatland bog through the Project's Payment for Ecosystems Services scheme.
5. According to the Cambrian Mountains Society, across the whole of the Cambrian range there are fifty recorded species of lichen, 300 species of mosses, thirty species of ferns, 450 species of flowering plants, 100 species of breeding birds, thirty-five species of mammals, thirty species of butterflies, twenty species of dragonflies and damsel-

Endnotes

flies, forty species of hoverflies, ten species of ladybirds and fifteen species of dung-beetles.

Chapter 2: Growing

6. These are the opening lines of the poem 'Reservoirs' in R. S. Thomas, *Collected Poems 1945–1990* (London: Phoenix, 1993), p. 194.
7. The poem was published in Ceiriog's collection *Oriau'r Hwyr* (Rhuthyn: I. Clarke, 1860), p. 30.
8. www.library.wales/discover-learn/digital-exhibitions/europeana-rise-of-literacy/poetry-volumes/oriaur-hwyr-sef-gweithiau-barddonol-john-ceiriog-hughes.
9. Available in Welsh with English translation, by Katie Gramich and Catherine Brennan, *Welsh Women's Poetry 1460–2001: An Anthology* (Aberystwyth: Honno, 2003).
10. From *Antigone*, cited in Hans Jonas, *The Imperative of Responsibility: In Search of an Ethics for the Technological Age* (Chicago: Chicago University Press, 1985), p. 2.

Chapter 3: Powering

11. The etymology of *gwlanen* derives potentially also from Middle English *flanneol* and Old French *flaine*.
12. From *An Historical and Archaeological Study of the Industrial Heritage of Newtown, Powys, Mid Wales* by Mark Walters (CPAT Report No. 552, 2003).
13. The trading name for the Going Green for a Living Land Trust.

Chapter 4: Bordering and crossing

14. Brian Walters, *Severn Tide* (Cirencester: Alan Sutton Publishing Ltd, 1987).
15. The Senedd has power to issue legislation in areas that include health and social care, housing, education, transport, business, economic development, social services, language and culture, the environment and local government. See: *www.centreonconstitutionalchange.ac.uk/the-basics/what-powers-does-senedd-welsh-parliament-have*.

16. '*Arglwydd, arwain trwy'r anialwch*' ('Lord, lead me through the wilderness').

17. The story is recorded in the Welsh oral tradition and in the sixteenth-century text *Historia Taliesin* (*The Tale of Taliesin*), retold in Miranda Aldhouse-Green and Ray Howell, *Celtic Wales* (Cardiff: University of Wales Press, 2017).

18. Rowan Williams and Gwyneth Lewis, *The Book of Taliesin: Poems of Warfare and Praise in an Enchanted Britain* (London: Penguin Classics, 2019).

19. *www.history.ox.ac.uk/article/how-england-became-the-sweetshop-of-europe*.

Chapter 5: Meandering

20. *www.britannica.com/video/185625/meanders-formation-rivers-streams-disturbances-disturbance-stream*.

21. Mary Webb, 'Vis Medicatrix Naturae', in *Poems and the Spring of Joy* (London: Jonathon Cape, 1935), pp. 127–8.

22. For further discussion: *www.nature.com/scitable/knowledge/library/overview-of-hominin-evolution-89010983*.

23. *www.geolsoc.org.uk/Geoscientist/Archive/October-2011/Charles-Lyell-and-deep-time*.

Chapter 6: Industrialising

24. Yasmin Anwar, 'Where do our minds wander? Brain waves can point the way', *www.news.berkeley.edu/2021/01/18/where-do-our-minds-wander-brain-waves-can-point-the-way/*.

25. Charcoal and coke produce less residue than wood or coal. Residues clog the furnace and can prevent the processes needed for iron-ore extraction from taking place.

26. For example, between 1990 and 2021, carbon dioxide, methane and nitrous oxide combined rose by almost 50 per cent, with carbon dioxide accounting for about 80 per cent of this increase: *wmo.int/news/media-centre/more-bad-news-planet-greenhouse-gas-levels-hit-new-highs*.

27. For further discussion: *www.un.org/en/climatechange/science/causes-effects-climate-change*.

28. *www.fossilplants.co.uk/adferiad-giving-back/*.

Endnotes

29. *www.huffingtonpost.co.uk/rupert-read/thrutopia-why-neither-dys_b_18372090.html.*

Chapter 7: Caring

30. *top-of-the-poops.org/waterway/severn-trent-water/river-severn.*
31. *www.greenpeace.org.uk/wp-content/uploads/2024/07/plastics_v08.pdf.*
32. For example, this ongoing issue was raised by Shropshire MP, Julia Buckley, in 2025, in a meeting with the Environment Agency. *www.shropshirelive.com/news/2025/02/22/crackdown-on-river-polluters-discussed-as-mp-meets-environment-agency/.*
33. *www.theconversation.com/why-the-uk-government-is-relaxing-rules-for-river-pollution-212505.*
34. For example, the UK 2024 Water (Special Measures) Bill gives public bodies and regulators, such as the Environment Agency and Ofwat, specific new powers to bring criminal charges against persistent lawbreakers, leading to imprisonment; to introduce severe and automatic fines for offences; to ban the payment of bonuses to executives of water companies not meeting required standards; and to ensure independent monitoring of every sewage overflow outlet. Wider measures in the Bill to strengthen regulation will include a new statutory requirement for water companies to publish annual pollution incident reduction plans, which set out steps they are taking to address their pollution incidents and prevent reoccurrence: *www.gov.uk/government/news/landmark-legislation-to-crack-down-on-bosses-for-polluting-water.*
35. *www.severnriverstrust.com.*
36. Numbers as of 2024. For updates and more details: *www.montwt.co.uk.*
37. Numbers as of 2024. For updates and more details: *www.shropshirewildlifetrust.org.uk.*
38. Other factors include: climate change, urbanisation, pollution, hydrological change and invasive non-native species woodland manage-ment. For further discussion: *www.naturalresources.wales/evidence-and-data/research-and-reports/state-of-natural-resources-interim-report-2019/challenges/.*

39. www.scientificamerican.com/article/only-60-years-of-farming-left-if-soil-degradation-continues.
40. www.theguardian.com/profile/david-r-montgomery.
41. www.ww3.rics.org/uk/en/journals/land-journal/making-the-change-to-regenerative-agriculture.html.
42. www.besjournals.onlinelibrary.wiley.com/doi/full/10.1111/1365-2664.14246.
43. For further discussion: www.news.climate.columbia.edu/2018/02/21/can-soil-help-combat-climate-change.
44. For example, read the interview with Carol Gilligan in 'Ethics of Care': www.ethicsofcare.org/carol-gilligan/.
See also Maria Puig de la Bellacasa, *Matters of Care: Speculative Ethics in More than Human Worlds*, 3rd edition (Minneapolis: University of Minnesota Press, 2017).
Joan Tronto and Berenice Fisher, 'Toward a Feminist Theory of Caring', in Emily Abel and Margaret Nelson (eds), *Circles of Care* (Albany, New York: State University of New York Press, 1990) pp. 36–54.

Chapter 8: Carving new routes

45. www.felinfach.com/pages/welsh-patagonia.
46. *Agor* is the Welsh word for 'open'. It has multiple significance as the name for this organisation, representing the opening of the Welsh landscape for extraction of resources, the use of modern technologies to open up records to a wider audience and the opening of minds to Welsh histories: www.agor.org.uk/cwm/themes/life/society/migration.asp.
47. www.gov.wales/welsh-language-data-annual-population-survey-april-2022-march-2023.
48. Hannah Arendt, 'Crisis in Education', in *Between Past and Future: Eight Exercises in Political Thought* (London: Penguin Classics, 2006, first published in 1961), p. 193.

Endnotes

49. This was a government-sponsored programme to encourage school-children and local councils to plant trees. It came at a time when the Dutch elm disease epidemic had devastated these trees in the UK and radically changed the landscape.
50. *www.teach.ocr.org.uk/naturalhistory.*

Chapter 9: Travelling onwards
51. *www.kemerton.org/wildlife_sites_upton_ham.htm.*
52. *www.caemabon.co.uk/single-post/2019/08/05/in-search-of-our-lammas-gods.*
53. An independent not-for-profit group of scientists and communicators who research our changing climate and how it affects people's lives.
54. *www.coastal.climatecentral.org/map/11/-2.1061/51.9254/?theme=sea_level_rise&map_type=year&basemap=roadmap&contiguous=true&elevation_model=best_available&forecast_year=2050&pathway=ssp3rcp70&percentile=p50&refresh=true&return_level=return_level_1&rl_model=gtsr&slr_model=ipcc_2021_.*

Chapter 10: Feeling the pull of the tide, the pull of home
55. *www.gloucester500.co.uk/brief-history-town.html.*
56. *www.gloucesterbid.uk/news/gloucester-docks-a-brief-history/.*
57. *www.butterfly-conservation.org/sites/default/files/2023-01/State%20of%20UK%20Butterflies%202022%20Report.pdf.*
58. Research conducted by the National Trust, World Wildlife Fund (WWF) and the Royal Society for the Protection of Birds (RSPB): *www.wwf.org.uk/press-release/save-our-wild-isles-campaign-launched.*
59. I draw here on a range of sources including: Pamela Petro, *The Long Field* (Bridport: Little Toller Books, 2021) and the BBC radio programme *Seriously Hiraeth, www.bbc.co.uk/programmes/p04kk8mg.*
60. *www.joebedford.co.uk/melissa-harrison/* and *www.monumenttotransformation.org/atlas-of-transformation/html/n/nostalgia/nostalgia-svetlana-boym.html.*

Chapter 11: Reuniting
61. *www.audubon.org/magazine/spring-2022/a-brief-history-how-scientists-have-learned-about.*

62. *www.education.nationalgeographic.org/resource/continental-drift/*.
63. *www.britannica.com/technology/cable-stayed-bridge*.

Reflecting
64. *www.artbasel.com/stories/otobong-nkanga-chooses-life*.